Improving Seed Conditioning

Improving Seed Conditioning

Bill Gregg, PhD

CRC Press is an imprint of the
Taylor & Francis Group, an **informa** business

CRC Press
Taylor & Francis Group
6000 Broken Sound Parkway NW, Suite 300
Boca Raton, FL 33487-2742

CRC Press is an imprint of Taylor & Francis Group, an Informa business

Printed on acid-free paper

International Standard Book Number-13: 978-1-138-03254-5 (Hardback)

Visit the Taylor & Francis Web site at
http://www.taylorandfrancis.com

and the CRC Press Web site at
http://www.crcpress.com

Printed and bound in the United States of America by
Edwards Brothers Malloy on sustainably sourced paper

Dedication

This publication, an effort to help the seed industry provide better service to farmers while earning reasonably from efficient performance, is respectfully dedicated to the memory of two persons who played an important role in my learning efficient and quality-oriented seed conditioning and in my dedication to helping seed conditioners improve their seed quality and performance efficiency.

Mr. Oliver Steele, Oliver Mfg. Co., *who taught me the importance of, and how to achieve, precise adjustment of quality seed conditioning machines.*

Mr. Lakshmi Sagar, Agri-Osaw Industries, *who taught me the necessity of having operators who understand and can make proper adjustments and integrate them to get precise operation of a machine.*

I want to acknowledge the immeasurable value of their contributions and influence to me and to the world's seed industry.

Contents

Acknowledgment

This publication should be of great value to seed conditioners. However, it has been a significant work to compile, prepare and organize the material included in the publication.

This collection of information and applications would never have seen the light of day without the huge amount of work, help and support provided by my wife, Orawan Chonlavorn Gregg. She has contributed a major effort in compiling, organizing and preparing the information herein.

I would like to acknowledge my wife's significant contribution to preparing this information and getting it into a form which will be useful to seed conditioners. We both hope that you will get much use for this material for many years!

Introduction

Improving farm productivity and the quality of farm produce are essential basic components of improving the health, well-being, and quality of life for people all over the world. One of the most critical factors required to achieve improved agricultural productivity is achieving and making generally available significant improvements in the physical, physiological, and genetic quality of seed that can plant and establish crops that produce higher yields of better produce and take less seed to do it. Seed conditioning is a critically essential part of creating better seed. And, the best way to maximize efficiency, economy, and accuracy in seed conditioning is to improve the skills and technical competence of seed conditioning operators and staff.

This manual was developed from experiences gained over lifetimes of working toward the betterment of seed conditioning, and it is intended to help implement the worldwide need to continue improving the quality of seed without major cost increases and to make good seed more readily available to more farmers worldwide.

This is an in-depth study guide prepared as an organized program and guide to help the individual seed conditioner and all the plant's staff, on their own initiative, to gain in-depth knowledge and experience to become expert professionals, getting the best conditioning and seed quality from the equipment in the shortest time and least cost.

Its structure also makes this program a complete guide for study and laboratory work sessions for organized classes on seed conditioning. It covers all needed areas—without skipping essential components—and includes all conditioning machines for individual machine study, allowing the studying operator to

- Select the machines—including brand and model—for which he/she needs increased competence
- Study when and as time permits and makes it possible for them to gain in-depth knowledge and understanding of the capabilities, adjustment, and requirements of machines as they are actually operated on seed lots
- Make concentrated studies on adjustment and operational performance of machines that are actually involved and information desired

It can also be used for organized class sessions, at any level. The format of this program allows it to be used with any models of any brand of machines.

It is intended to improve the competence, efficiency, and economy of efforts to use seed conditioning to maximize seed quality at minimum cost and effort, to help make this a better world by providing farmers with ever-better seed at the same or lower price!

To meet this need

This technical assistance program has been developed for use either as an individual self-study program or as a text for organized classes. Its objective is to help resolve the staff technical competence problem in a different way: (1) it provides a form of training that will help conditioners learn how to adjust and use machines at their full capacity and separating capability; (2) it is in a format such that unemployed students can take this study course in full-time technical or university courses; and (3) it is also in a format that provides conditioners who are working full-time to be able to take this training as *intensively guided self-study* to improve their competence and value to their employer so they will not become technically obsolete and will improve their competence and value to their employers while they work! Recognizing this need, this "train by learning to do it properly" self-teaching program and manual was created to enable the operator to

1. Study each machine in-depth, what it does, and how it does it
2. Learn to make adjustments so as to get the best possible separation at the best possible operating capacity with the least possible loss of good seed

This training manual can be used as

1. A detailed text for organized classes at the level of university or special technical development programs
2. A professional self-study guide for personal development while working in seed conditioning to maximize professional development for conditioning managers, operators, and crop/seed specialists

It is designed so that the owner of this manual can work with and on this basic manual so that he/she can constantly gain a new and in-depth understanding of the precise separator models he/she works with so as to understand them precisely and continue honing his/her skills and achieve maximum precision and efficiency as machines, operating conditions, and crop seed change. Recording this information in his/her personal manual will reflect his/her growing experience and knowledge in his/her constantly growing personal conditioning atlas.

> This study program should be required professional development and work performance and improvement approach of all staff in every seed program of government or private sector.
>
> **—Executive of an International Seed Company**

How this program works

A major problem that is generally not recognized or considered in the seed industry of most countries is a lack of the skills of program managers and operating personnel to get the maximum precision, capacity, and separation from the conditioning machines installed in the seed conditioning plant. The most effective and least-costly way to increase separating precision and capacity, while reducing loss of good seed, is to give both management and operating staff in-depth skills to get the most out of each machine.

However, there has been no program that could supply the in-depth training needed. Even if such training had been available, it would have been focused on one set of conditioning machines of specific brands, thereby leaving untrained all personnel in those plants where other brands or models of machines were installed.

This program is designed to improve seed conditioning operation and management skills in order to enable staff to get the maximum performance out of the machines in their own plant so as to reduce time, cost, and seed losses and to improve seed quality, output of good seed, and timely performance of seed conditioning.

This program is designed so that the equipment actually installed in each person's plant is used in the training. And, when different equipment is installed or a person moves to another plant with different equipment, the training can be adapted to fit the new conditions. This training is based on the actual equipment available.

Getting maximum benefit from this technical assistance program

1. This program is a novel approach to technical assistance, transfer of technology, training of staff, and development of local technology-applying modern technical development based on intensive study and training on the equipment available supplied by the technical/industrial/economic development program. It is also focused on developing self-reliance of local program staff and exposing them to methods of technology assimilation by on-site and self-stimulating activities, under actual local operating conditions.
2. Since most staff in existing plants need this training but cannot take leave long enough to take a formal training course, this training program is designed in a format that not only allows it to be used in formal training courses at the university or industry level, but also by one (or more) individuals via the self-study materials in either off-season training or training as seed are being conditioned in the regular working season.
3. This training material can be used for any model or brand of seed conditioning equipment. Even as machines are replaced with other models or brands, or the specialist goes to another plant, this program will still be useful.
4. To minimize confusion and help focus attention, much of the material is organized in question form, or in a form that directs attention specifically toward the aspect being considered.
5. The lessons consist of two parts: (1) the question or statement identifying the aspect being considered on the machine and (2) the Conditioner's findings from individually conducted operations on his/her specific machine in this aspect.
6. Different people would often word things differently, or may have more or less in discussing each aspect. To accommodate this, each lesson is divided into two parts: (1) a listing of questions/aspects being considered at the time and (2) then blank space for the Conditioner to write down a complete description of his/her findings. To handle this approach, the program is organized into a loose-leaf binder format. Thus, for each machine or topic, a list of questions and points for consideration is provided in a loose-leaf notebook. When the Conditioner studies these points, he/she writes down his/her findings on locally available, loose-leaf paper and inserts it in the notebook after the pages that include each question or point. This enables each Conditioner to include more or less in his/her personal notebook.
7. This is done while the Conditioner is making an initial study for training. However, he/she builds a complete and personal volume of information that can be used as a reference at any time in the future when a similar conditioning situation arises.

8. When a new or different machine is involved, the Conditioner can simply remove the old self-prepared information and do a new study of the new machine and then prepare his/her new set of answers/information on the new machine.
9. A large volume of information will normally be developed as a result of this study. Therefore, this training material is divided into two separate loose-leaf binders to facilitate handling the material.
10. Each Conditioner should have his/her personal copy of this training material and keep it for future reference when similar situations are encountered. This should become a guide and ready reference for the Conditioner's entire remaining work life.

Making a personal permanent reference notebook

As his/her studies progress, the Conditioner cannot possibly remember the critically essential details of every activity.

The Conditioner should use this training program as a means of preparing a *Permanent Personal Reference Notebook* (PPRN), as the personal property of the Conditioner, that can be consulted at any time. The Conditioner should take copious and complete notes during the appointed study periods (either lectures, demonstrations of applying a specific condition, or personal study applied during the actual conditioning of seed lots in the Conditioner's commercial seed conditioning plant).

Start the PPRN with objectives and personal information:

- Conditioner name.
- Date.
- How the training is being implemented.
- Seed conditioning plant/machine models, size, and location.
- Description of installation and feeding seed/waste materials to/from the machines. Any problems encountered with operation of the machine, or points that need to be discussed with the training leader, to clear up any problems. After initiating the training program, each Conditioner, on his/her own personal initiative, should take complete notes on each topic and keep them in an organized form in his/her personal training manual. This has the added advantage of providing full information on points that are of special interest to each Conditioner.

Procedure: After studying seed conditioning principles and basic operations and equipment of each detailed aspect required to condition the seed, answer the questions listed herein for each machine/operation/condition, in a manner that provides complete information and can be referred to for guidance in future operations. Carefully and completely record all new information in your manuals.

Using this training program

These analyses were designed for easy use by either of two approaches:

1. Individual study by persons actually working in conditioning plants, accompanied by study of detailed information on their own machines and models—specifications, operating and repair manuals, etc.
2. In-plant laboratory sessions of organized technical or university-level classes.

The conditioner–studier, in either case, should obtain several loose-leaf notebooks and a supply of paper punched to fit into the notebooks. The appropriate sections in this study guide should be placed in a notebook, as the sections are studied. Additional information such as diagrams/pictures of the machine and different components, adjustment and maintenance guides for the machine, operator manuals, special information on the machine, and so on should be included in the notebook. The studier should end up with a personally prepared in-depth efficient operating guide on each machine model that is in his/her specific plant, as part of a complete set of detailed guides for all machines. In formal class use, limited time usually does not permit full completion of these analyses during the class. Complete the necessary work during class periods, and complete certain portions of the analyses during individual study time. Confer with the instructor as necessary and participate in class discussions and activities. The instructor should also prepare sufficient copies of diagrams, machine manuals, and so on to provide to all participants. Complete each machine analysis only after the class covers the machine and/or after a thorough personal study of technical material, descriptive machine literature, and other available information. Before in-plant laboratory analyses, some parts of the analysis of machine characteristics can be completed individually using information from lectures, individual study, information provided to participants, etc. During in-plant detailed "laboratory analysis" of a machine, complete remaining parts of the machine characteristics questions by detailed study of the machine and its actual operation. Where analysis of samples is required, these may be done later and the study analysis then completed; or, several persons may work together. Detailed study of some machines may not be desired; omit analyses for these machines. These analyses were planned according to each machine's characteristics of operation for flexibility in emphasis on different machines to meet different needs.

Updating or adapting this manual and your personal reference notebook

This manual was designed for the entire seed industry and may not be as complete as desired for some machines. In such cases, the instructor or participant should add additional questions and discussions as required to make the study as in-depth as desired for specific machines. The participant should build his/her own detailed manual to fit the machines, models, and needs of their individual situation and the machines in it.

Caution: Operating machines are dangerous! Use extreme care to keep hands, clothing, etc., away from moving parts!

The operator makes seed conditioning separators operate better

The role and objectives of seed conditioning are as follows:

1. Make all seed lots of quality equal to or higher than the required standards
2. Complete the work and have seed ready for marketing in the least time possible as farmers need the seed
3. Minimize operating costs
4. Maximize per-hour throughput
5. Minimize damage (breaking, etc.) to seed

6. Minimize loss of good seed in the reject fractions
7. Maximize removal of undesirable seed and materials

The most important tool in seed conditioning and the performance of each machine is the *operator.*

The operator decides what needs to be done, selects the machines to use, puts the machines in an effective sequence, adjusts them to achieve the best separation, determines the best operating capacity of the machine and the entire sequence, and determines how much good seed is lost and whether the separation is acceptable.

To do this and get the most efficient operation, fastest output, lowest cost, minimum loss of good seed, best separation, and best seed quality require in-depth knowledge of the machines and what they do and how they do it—and knowledge of the seed!

This usually requires in-depth training and regular upgrade training, as well as years of in-depth study and expertly supervised practical experience. These are generally not available to most conditioners. How can this information be developed and transferred to the operators who need it?

This study guide was prepared to provide conditioners with an organized approach to learning how to get the full capabilities of seed conditioning machines. This is an extension and applied approach intended to build on and develop practical application to increase utility of basic information provided in the special texts, *Seed Conditioning* by Bill Gregg and Gary Billups, Volumes 1, 2, and 3 (Science Publishers 2010).

This is an organized, guided approach to help identify, know, and understand each factor of each machine and to observe the effects of changed adjustment on the seed.

It is presented in the format of analyses to give the operator a guided study program to gain in-depth knowledge of his/her specific in-plant model of each conditioning machine and to gain some guided experience in its operation. Thorough study of one model will enable a person to make a similar evaluation of other models as necessary and familiarize himself/herself with their operation in a short time.

This guide can be used by

1. Individuals working in seed conditioning who want to improve their performance
2. Organized classes or groups studying aspects of seed conditioning

This guide can benefit

1. Seed conditioning staff who want to improve their proficiency, performance, and efficiency
2. Seed enterprise staff who want to enhance quality control, marketing, production, and management so that they can cooperate more effectively with conditioning
3. University seed technology and engineering courses laboratory sessions where seed conditioning is studied; this guide can be used as the laboratory program for two consecutive semesters of lab periods
4. Vocational training courses to create a pool of competent staff for seed enterprises
5. Short-course training sessions for seed industry personnel
6. University agricultural and engineering research, teaching, and extension personnel
7. Equipment supply, sales, maintenance, and development staff

Study of the machine's characteristics draws attention to important parts of the machine, their functions, how they are used, and how the machine is used in seed conditioning

Experience in operating the machine is to demonstrate what the machine does to a seed lot, how it does it, how each control affects the machine's performance, and important aspects of its operation.

Analysis of a machine and its operation in detail is necessary to ensure sufficient competence to understand the machine and use it successfully.

Conscientious completion of these analyses will not make a person an experienced conditioning specialist. However, he/she will have the basic knowledge and understanding to readily develop the desired proficiency.

chapter one

What seed conditioning is and does

The conditioner should prepare a complete answer to each indicated question (and other questions that occur to the conditioner) and place them in his/her notebook.

1.A Importance of seed

Seed is the basis of the life of plants, and farmers plant seed to produce the crops that generate their incomes. The better the quality of the seed, the greater is the production grown from the seed, and the higher is the farmers' incomes.

To initiate the training, and form the conditioner's basic concept of WHY CONDITION SEED, discuss the following points and use them as the starting reasons for seed conditioning. Also use them to start your personal reference book.

1.A.1 Explain the role of seed in agriculture and food production and why seed quality is important.

1.A.2 Explain the meaning of seed quality.

1.B Seed conditioning

Seed conditioning uses specific machines, each of which performs a particular operation in the series of activities of seed conditioning. When all combined and properly executed, these activities can significantly improve the quality, vigor, genetic capability, and production capacity of the seed that farmers use to produce crops. Seed conditioning is considered to be the first step in improving the productivity and income of farmers.

1.B.1 Explain the meaning of seed conditioning.

1.B.2 Explain how seed conditioning improves seed quality.

1.B.3 List the components of seed quality and explain why seed quality is important.

1.B.4 Explain where seed conditioning fits into the sequence of operations that produce seed; handle and store it; and deliver it to farmers who produce the crops of food, feed, and fiber that are necessary for people to have good lives.

1.B.5 Does seed quality occur naturally? Is it a permanent condition? Explain.

1.C Conditioning plant facilities

The following facilities (and operations) should be provided in a complete and self-supporting seed conditioning plant:

a. Truck scales to weigh incoming and outgoing trucks
b. Truck parking, inspecting loads, servicing, and handling
c. Management and administrative offices
d. Internal quality control
e. Raw seed receiving facilities
f. Seed drying
g. Raw seed storage
h. Conditioning
i. Waste (screenings) handling and disposal

j. Conditioned seed storage
k. Conditioned seed and loading facilities
l. Workshop

1.C.1 *Describe the work and use of each of the above-mentioned Seed Center conditioning components.*

chapter two

Seed conditioning plant staff and contract growers

2.A Relationships and interactions of plant staff with contract seed growers

The improved seed supplied to farmers is, in most cases, produced by specially trained farmers. These farmers are contracted to be seed producers and produce high-quality crops that will be conditioned and sold to other farmers for use as high-quality seed to produce their commercial crops.

After harvest, the seed of high genetic and varietal purity that is produced under special conditions by the contract growers is taken to the seed conditioning plant. Here, the seed is processed through a series of highly specialized machines that use mechanical systems to remove undesirable seed/particles. This operation is done to improve the seed's quality (i.e., physical purity, germination, vigor) before it is supplied to farmers who plant this improved seed to ultimately produce better crops to be sold on the market. The staff of the seed conditioning center are deeply involved in all production activities and advise contract growers on every operation, to be sure the seed produced are of the maximum quality possible.

2.A.1 Describe the relationship between contract seed growers and seed conditioning staff.

2.B Selecting contract seed growers

The better farmers in an area are selected to be contract seed growers, since they already understand and apply many of the practices that are required to produce high-quality crops. Seed conditioning staff work with local agricultural Extension, research staff, and farmers; thus, they usually already know the local farmers who would produce the best crops and consequently make the best seed growers. In addition, seed conditioning staff already know the farmers who are more likely to understand and apply the best crop production practices to ensure higher quality and higher yields of the contracted seed crops. Involving seed conditioning staff in selecting, training, guiding, and advising contract seed growers speeds up the process of grower selection and guidance and ensures that farmers know and trust the people who will guide and advise them.

2.B.1 *Why should the best local farmers be selected to produce contracted seed crops?*

2.B.2 *Why is it wise to involve local seed conditioning staff in selecting, training, and guiding farmer contract seed growers?*

2.C *Training, supporting, and supervising contract growers*

Seed conditioning staff have special training and experience that enable them to understand and anticipate the local conditions that affect seed crop yield and seed quality. These conditions are many and wide ranging (e.g., field fertility, weed and insects problems, farmer experience, equipment, facilities, and location), and they may occur at any stage of crop growth. These conditioning staff are also already in the area, so they are familiar with its conditions and problems, and they know many of the local farmers.

Seed conditioning staff thus are trained in good crop production and are able to recognize field conditions that affect seed yield and quality BEFORE they create bad conditions that cannot be corrected. They can help farmers avoid problems and produce maximum yields of good-quality seed. The conditioning plant staff are thus good farmer advisors who are already in the local area and know local conditions, farmers, potential problems, and effective preventive practices.

This knowledge not only helps the contract growers produce more and better seed but also helps the seed conditioning staff be aware of likely or actual problems and what to look for when the seed crops are harvested and delivered to the conditioning plant.

Close cooperation with contract farmer growers also helps the conditioning plant staff to know in advance the conditions, quality, and requirements of the seed before harvest, so they can make adequate preparations for handling the seed when it arrives.

2.C.1 *Why should seed conditioning plant staff guide and advise contract seed growers who will produce the plant's seed?*

2.C.2 *How does working with contract growers help the conditioning plant staff know conditions and quality of seed that they will receive and be able to prepare in advance to make the best-quality seed at the lowest cost?*

2.C.3 *How often, and for what, should Seed Plant staff be in contact with contract growers?*

2.D Advising contract growers on quality control

Seed Plant staff have training in seed quality achievement and maintenance, and they are ideally qualified to guide farmer contract growers on how to identify materials that lower seed quality, recognize potential problems, and correct them before they affect seed yield or quality.

2.D.1 *What are some of the potential field factors that affect seed yield or quality that conditioning plant staff could recognize and help contract seed growers correct before they damage seed quality?*

2.E Advising contract growers on harvesting and handling seed

Harvesting and handling procedures are extremely critical to maintaining the quality of seed. Seed must be at the proper stage of maturity when they are harvested; handled in specific manners to prevent loss of seed quality; and delivered to the conditioning plant effectively to prevent damage by, for example, seed moisture content, insect damage, and contamination. Conditioning plant staff know these conditions and the facilities that are available. They can advise farmers on the ideal stage for harvest and help ensure that all equipment, bags, and other relevant materials are cleaned and will not cause contamination. They can advise farmers on on-farm seed storage and handling and arrange delivery to the seed conditioning plant.

Delivery and maintenance of high seed quality are not simple. They involve cleaning out of harvest, handling, and bagging equipment; using proper harvest methods at the best stage of maturity; operating harvest so as to protect the seed and prevent damage or contamination; determining seed moisture content and drying needs; protecting seed quality until the seed is delivered to the conditioning plant; arranging proper reception and handling in the Seed Plant; maintaining proper cleanliness and identity; and preventing admixture at all stages.

2.E.1 *List some of the things plant staff can advise contract growers on to ensure timely completion of quality-ensuring production operations, harvest, and handling.*

2.E.2 *Why are inspectors from the conditioning plant essential advisors and guides for contract farmer seed growers?*

2.F *Delivering raw seed to the conditioning plant*

When plant staff work with contract growers, they know the condition of the seed (e.g., maturity and quality at harvest; moisture content; and need for drying, insect infestation, conditioning requirements, and seed treatment). They are thus able to guide growers in proper harvest methods at the ideal stage; prepare effectively for receiving the seed; schedule seed delivery; and prepare handling, storage, and conditioning so that seed quality is protected and operations are successful, timely, and efficient. Seed Plant advisors should guide harvest at the proper stage and ensure immediate delivery of harvested seed to the conditioning plant, where it can be prepared properly and handled safely.

2.F.1 *List some of the preparations for receiving freshly harvested seed that need to be done before the seed arrives at the conditioning plant, and how plant staff working with contract growers can help plant staff to be ready to handle the arriving seed efficiently and safely.*

2.F.2 *What are the equipment and facility requirements for receiving raw seed at the conditioning plant?*

chapter three

Seed conditioning plant location

Several important factors must be considered in selecting the location for a seed conditioning plant. These factors affect not only the quality of seed but also the cost of operations and the relative ease with which the operations can be conducted.

3.A Near the production area

The plant should be located near the area where the contracted farmers produce the seed. This proximity makes it more efficient and timely for seed staff to work with and assist the seed growers. It is also more efficient to deliver raw seed to the plant for drying, because drying must be done properly and immediately after harvest. Similarly, the material that is removed during conditioning does not have to be transported for long distances, thereby reducing operating costs.

The seed conditioning plant should be located as near as possible to the seed producers and contract growers. This nearness makes it easier and more efficient for supervising personnel to contact and assist growers and for growers to contact the seed plant staff. It also expedites movement of supplies to growers and reduces cost and time required to transport harvested seed from growers to the conditioning plant. It also helps ensure delivery of seed to the plant immediately after harvest and thus helps prevent loss of seed quality because of lack of safe storage or drying facilities on the farm. Nearby location also reduces costs of inspection and assistance to growers, and increased frequency of contacts helps maintain good relations between the seed program and growers.

An often-overlooked factor is that along with the good seed, contaminant materials that are removed during conditioning are also transported from the grower to the conditioning plant. If 5%–10% of the raw seed must be removed in conditioning, added distance to the plant increases the cost of delivering waste materials that are of no value and must be removed in conditioning.

3.A.1 Explain why the seed conditioning plant should be located near the contract growers who produce the seed.

3.A.2 Describe the reasons for some of the visits to contract growers by seed center personnel.

3.B Not near trash, weeds, etc., which attract rats and other pests

Locating the plant in a "clean" area helps reduce damage caused by rats and other pests and reduces control costs. If the area outside but around the plant is trashy or overgrown, it will attract and provide cover to birds, rats, and various other pests that will unavoidably enter the conditioning plant and its area and damage seed. The entire area around the plant should be clean and free of trash, weeds, brush, and so on to help ensure that seed-damaging pests will be minimized. Experience has shown that locating the plant in a clean area will significantly reduce entry of insects, rats, and other pests into the conditioning plant, thereby reducing damage to seed and reducing control costs.

3.B.1 *Why is it important to have a site and surrounding area that are free of trash, weeds, etc.?*

3.B.2 *How can you reduce rat/insect attraction to seed center lands?*

3.C Dry, low-moisture area with good drainage

Moisture content of seed must be kept low to prevent loss of germination and vigor. The first requirement is to locate the Seed Center in a dry area because seed is hygroscopic and absorbs moisture from damp air.

The seed conditioning plant area MUST be kept clean and dry. Furthermore, the entire surrounding area should be clean and dry; if it is not, the dampness and humidity around the outside of the plant will result in high humidity in the plant area that will unavoidably have a damaging influence on seed quality. Seed is hygroscopic and absorbs moisture from high-humidity air. The entire surrounding area should be clean and devoid of trash/weeds/insect and rat cover and not have heavy traffic and "people activity." The entire area must have good drainage, especially in rainy seasons.

3.C.1 *Why is it important to have a dry, well-drained area for the seed conditioning plant?*

3.C.2 *How is moisture and high humidity in the surrounding area transferred to seed?*

3.C.3 Discuss the effects of high moisture and humidity on seed and seed storage.

3.D Minimal traffic and "outside" personnel

If there are a lot of people and vehicles that enter or come around the Seed Center, there is a greater likelihood of problems developing. A quiet, out-of-the-way location is thus ideal because it promotes fewer problems and less effort and cost in control measures.

3.D.1 Describe a good location for a Seed Center, in terms of traffic. What are some of the potential problems, and how can they be reduced?

3.E Ready access for trucks carrying seed in and out

Despite efforts to minimize traffic in and around the Seed Center, all the seed is trucked in for conditioning and then trucked out to the dealers and farmers who need the seed. Roads to the Seed Center must be in good condition and able to handle the maximum seasonal traffic expected.

Roads from growers to the conditioning plant should be in good condition and passable at all times. This facilitates moving raw seed from grower farms to the conditioning plant where the seed can be properly dried, protected from rats and insects, and handled and introduced into the good seed program safely and quickly after harvest.

Although all raw seed must be delivered to the plant, the cleaned seed must again be transported to markets where it is delivered to farmers who use the seed. Commonly, smaller quantities are delivered to numerous locations for marketing conditioned seed. This approach contrasts with that of the delivery of raw seed, where large quantities are delivered to one location—the conditioning plant.

3.E.1 Describe the traffic and seasonal nature of movement of seed at the Seed Center.

3.E.2 Why are good roads between the growers and the conditioning plant important?

3.E.3 *Why are good roads between the conditioning plant and the markets for the clean seed important?*

3.E.4 *Describe the transport of freshly harvested seed from growers to the Seed Center. Why is it essential to deliver seed immediately after harvest?*

3.E.5 *Discuss the impact and effects of delays in delivering seed to the retail dealers and farmers who will plant the seed.*

3.F *Not in the immediate area or adjoining areas where there is much traffic*

A major problem in crowded suburban areas is the amount of vehicle and pedestrian traffic in or through certain areas. Such traffic around the conditioning plant will unavoidably result in people not involved in the seed program coming into the seed plant area. This type of traffic results in loss of good seed (as well as equipment and supplies) and contamination and movement of trash and foreign materials into the seed plant area. For safety of seed quality, plant personnel, equipment and facilities, outside vehicles, all pedestrians (adults and children), and animals must be kept out of the seed plant area. The plant area should be surrounded by a rat/pest/pedestrian-proof concrete wall; entries should be minimized, and guards should be stationed so they can keep out unauthorized personnel.

3.F.1 *List some of the problems that could arise because of excessive or undesirable traffic (both vehicular and people) in the area around the Seed Center.*

3.F.2 *Why must the status of the area be considered in all stages of the design and a seed conditioning facility be constructed in an area that can successfully operate in a local area?*

3.G Not near areas where seed plant noise and 24-hour/day operations would be objectionable

The seed conditioning plant must be in an area where its essential activities can be conducted when and as needed, without having to worry about or meet schedules of other activities in the adjoining area. Noise and dust from the seed conditioning operations are usually quite objectionable in areas where people live, or where there are retail operations such as stores and restaurants where there are usually considerable numbers of people.

3.G.1 *Why is it important that seed conditioning plant activities be conducted when, how, and as needed for the seed, rather than be influenced by adjoining activities or facilities?*

chapter four

Seed conditioning plant area

4.A *Adequate space for all required plant operations*

This requirement is a no-brainer: adequate space must be provided for each operation to be efficient. As volume grows, space and equipment needs should be compared with the actual situation, at the end of each conditioning season.

4.A.1 *How much space is required for each operation: receiving, drying, raw seed storage, conditioning, clean seed storage, shipping, and quality control?*

4.B *Concrete wall surrounding the entire area*

A solid concrete wall around the entire Seed Center site helps keep out rats, pests, insects, etc., that would otherwise be attracted to the seed held in the Seed Center. With a solid concrete wall, these pests can enter only through the two gates, and this entry can be largely controlled.

4.B.1 *Why is a solid concrete wall around the Seed Center worth its cost?*

4.C *Pave entire plant area with smooth-finish reinforced concrete*

If the entire area within the Seed Center is paved with smooth-finish concrete, the area can be used in busy seasons for operations such as temporary storage and drying.

4.C.1 *How can the smooth-paved open area inside the Seed Center be used when harvest, conditioning, shipping, etc., are at "full swing?"*

4.D Only two gates into the plant area

Two gates are all that are needed to provide full required access to the Seed Center. More than two gates create unnecessary risks and would not really help with operations.

These two gates should be located close enough together so as to allow one gatekeeper's booth and one gatekeeper/guard to control entry and exit at both gates.

4.D.1 *Why should gates and openings into the Seed Center be minimized?*

chapter five

Essential utilities

The Seed Center involves operations and people that create needs for certain utilities and services. These utilities can sometimes be obtained from local utility operations; however, the Seed Center itself must often make provisions to provide its own utilities.

5.A Fire protection

Buildings and facilities of the seed conditioning center must be constructed of materials installed in such a manner that risk of fire is minimized. Electricity supply and electrical equipment used should be of types and installations that are as fire resistant as possible. Smoking should not be allowed within the Seed Center. Occasional operations involving fire or fire danger (e.g., burning trash, welding, using gasoline for cleaning) must be avoided to the maximum extent possible. When such operations are unavoidable, use all possible precautions to minimize fire risk and have fire extinguishers and control facilities ready for use at the site.

Information on local fire departments should be shown throughout the seed conditioning center and their suggestions adopted. Arrangements should also be made to ensure that if or when needed, local firefighting/control agencies are quickly available for the Seed Center. In addition, local fire officials should be regularly invited to inspect the Seed Center and make recommendations on incorporating the best possible fire prevention program in the Seed Center.

5.A.1 *Discuss the components of a complete fire prevention system and how the support and participation of local fire control agencies can be obtained.*

5.A.2 *Discuss how fire prevention techniques can be incorporated into the Seed Center's daily operations.*

5.A.3 *Describe fire prevention/control facilities that should be installed in the Seed Center, e.g., an elevated water tank, sprinkler systems, fire-proof waste collection, and disposal.*

5.A.4 *What regular training, reminders, and procedures should be conducted in the Seed Center to emphasize constant attention to fire prevention?*

5.B Clean water supply

The Seed Center needs water for a range of operations such as human consumption, seed treatment, seed testing, and cleaning operating areas and trucks. Adequate pure clean water must be made available, either by connecting to local water supply systems or by installing adequate wells and water supply facilities within the Seed Center.

5.B.1 *Describe the water use of a Seed Center.*

5.B.2 *Describe the components of a clean water supply system for a Seed Center.*

5.C Adequate and dependable electricity

Most operations within a Seed Center require power, normally in the form of electricity, for motors, lights, and temperature control. The most economic source of electricity is usually a connection to the local electricity supply network.

Because of the timeliness required for some operations, the Seed Center may find it necessary to operate electricity generators at times to conduct essential operations in a timely manner required to maintain seed quality and to meet market requirements.

Motors and equipment in the Seed Center should be of the best fireproof category available. The electrical installation should be of the best industrial safety category, both to prevent fires and damage and to protect personnel and machines.

5.C.1 *Describe electricity needs and uses in a Seed Center.*

5.D Sewage

Sewage generated by operations and personnel must be disposed of in a safe and sanitary way. If connections to a local sewage system are available, they can be economically used. If a suitable local system is not available, the Seed Center must install its own system with adequate facilities, piping, septic tank, and disposal system such as disposal field.

5.D.1 *Describe the sewage disposal requirements of a Seed Center, and how the Seed Center can ensure that its system is effective and protects personnel and seed.*

5.E Safety and security

Safety of personnel, buildings, equipment, vehicles, and seed is a constant and extensive requirement in a Seed Center. Operating safety aspects must be identified and regularly emphasized through training sessions and reminders to staff. Outside persons should be kept out of the seed center as much as possible and constantly accompanied and guided when visitors are in the plant. Watchmen, walls, and other safety provisions must be used to prevent criminal problems.

During peak seasons, the seed conditioning operations usually operate 24 hours a day. However, at all times of the year, full-time 24-hour-a-day security personnel should be employed. A suitable wall (solid wall of smooth-finished reinforced concrete) around the Seed Center helps keep out suspect persons and rats and other pests. An enclosed guard post should be constructed and adequately staffed at the entry gate to help authorized personnel and keep out unauthorized persons. As are security personnel in other local operations, the Seed Center's security officers should be appropriately armed.

5.E.1 *Describe a complete safety and security program for a Seed Center to minimize injury and health problems of staff and visitors and to prevent theft, robbery, and other types of criminal problems.*

5.E.2 *During peak periods, Seed Center operations run 24 hours a day. Year-round, the Seed Center's security system should be active 24 hours per day. Explain why this level of security is important and how it can be achieved.*

5.E.3 *How can a good wall around the Seed Center site keep out rats and other pests?*

5.F Trash and garbage

The Seed Center should have access to local trash and garbage handling and removal programs that remove "normal" trash and garbage. However, the waste products generated in seed conditioning are usually handled separately by the Seed Center because of the volume and nature of the products. Trash, including generated waste products from conditioning, should be regularly removed and safely disposed.

5.F.1 *Describe why wastes removed by conditioning machines are different from "usual" garbage or wastes.*

chapter six

Conditioning plant facilities

A seed conditioning plant requires specific facilities to enable efficient performance of the essential activities.

6.A *Truck scales to weigh incoming and outgoing trucks*

Truck transport is required to bring in raw seed from producers and to deliver conditioned seed to dealers, farmers, or both who need the seed to produce their commercial crops. Truck scales are required to measure the quantities of seed coming in or going out.

6.B *Truck parking, inspecting loads, servicing, and handling*

Facilities are required to enable efficient handling of trucks and their loads.

6.C *Management and administrative offices*

Adequate offices and operating facilities are required for management of operations.

6.D *Internal quality control*

Adequate and efficient facilities are required to ensure that seed quality is always and constantly maintained at the required levels. Meeting this need involves operational supervision; examination and maintenance of facilities and storage conditions; seed sampling; seed testing; and modification of facilities, operations, or both as required.

6.E *Raw seed receiving facilities*

Incoming raw seed must be received efficiently so that there is no delay and no condition that would affect their quality or identity.

6.F *Seed drying*

Seed moisture must be held at safe levels at all times.

6.G *Raw seed storage*

Raw seed must be stored under conditions that ensure and protect its quality, ensure its identity, and enable efficient operations.

6.H *Conditioning*

Conditioning facilities, location, and operations must achieve desired seed quality, maintain seed purity and identity, and be completed at low cost and within the required time frame.

6.I Waste (screenings) handling and disposal

Conditioning generates waste products. These wastes must be handled and disposed of in a cost-effective manner that does not risk seed quality, interfere with operations, or generate insect or other pest problems.

6.J Conditioned seed storage

Conditioned seed must have storage that protects its quality, purity, and identity and allows efficient movement into marketing channels.

6.K Conditioned seed and loading facilities

Conditioned seed must be handled in a manner that ensures efficient loading of trucks and other aspects of moving the seed into the market.

6.L Workshop

Conditioning involves machinery and trucks that must be kept in good operating condition and be ready for use when they are needed. It is most efficient for the Seed Center to maintain its own workshop facilities and staff who are capable of handling the major part of maintenance and repairs needed in day-to-day operations.

chapter seven

Raw (nonconditioned) seed

7.A Receiving raw seed

A seed conditioning plant must receive raw (nonconditioned, or uncleaned) seed as it is harvested by contract growers and then refine the seed until it consists solely of good crop seed that are mature, undamaged, and alive and that when planted are capable of producing healthy vigorous plants of the desired crop and variety. Every operation within the plant must be designed and implemented in a manner that maintains the purity and condition of the seed, without damage or weakening.

7.A.1 Describe the harvest and delivery of raw seed and its receipt at the Seed Center.

7.B Delivery handling systems for raw (nonconditioned) seed

Different systems or conditions of delivering raw seed from contract growers to the conditioning plant are used: bulk seed in trucks, seed in bags, or seed in toteboxes handled by forklift. Each system has specific requirements for handling facilities, handling drying, storage, and introducing seed into the conditioning machine "line."

The typical modern seed conditioning plant can handle raw seed delivered by each of these three handling systems. The plant must have the space and the equipment required to handle the delivery systems used. For example, handling seed in bags requires forklifts and pallets, and handling seed in forklift toteboxes requires an adequate number of toteboxes and forklifts that are capable of handling the weight and size of the toteboxes. Handling bulk seed requires a truck dumping system, a truck drive-through the building where seed is received, an adequate-capacity receiving hopper installed beneath the floor and adequate provision for trucks to drive over the receiving hopper and to dump their loads of seed, a system of conveyors, an elevator and spouts adequate to take seed from the receiving hopper to appropriate and adequate bulk storage bins in sufficient numbers, and ability to handle the required capacity and number of different lots that must be handled and stored separately. The entire system must be self-cleaning so that different lots can be handled without delays required for cleaning out remaining seed left from the previous lot handled.

7.B.1 Describe each of the three systems of handling seed, and the equipment required to implement the methods.

7.B.2 Discuss the number and capacity of bins required.

7.B.3 Describe the self-cleaning aspects of the conveying, elevating, and spouting system and the bins that are required.

7.B.4 Discuss how the receiving bins are installed in the raw seed storage in relation to the receiving facilities.

7.B.5 Describe a complete seed labeling and record system.

7.C Methods of receiving raw seed

There are three different handling systems in common use for receiving incoming raw seed. Most Seed Centers use a combination of all three methods. The Seed Center must have the layout, equipment, and facilities to handle incoming raw seed safely, regardless of how it is delivered to the Seed Center.

7.C.1 Describe the different methods of receiving raw seed, the equipment and building design needed for each system, and how seed can be handled.

7.D Bulk in loaded trucks

This method is more commonly used for types of seed that are handled in larger volumes. To receive bulk seed, the Seed Center must have a receiving hopper of adequate size, installed inside the building, beneath the floor of the conditioning/storage building, so that the system is protected from the outside weather. Trucks must be able to drive into the building, cross over the receiving hopper (a grill of adequate strength must be installed over the hopper), and then dump their loads of seed into the receiving hopper. The hopper then feeds the seed into the receiving self cleaning elevator that delivers the seed to the desired place. This system requires an adequate number of bulk storage bins to receive incoming raw seed of different lots, varieties, and crops and a complete elevating and conveying system. If drying is required, the drying system must also be fed, either by (1) returning the seed from the raw seed storage bin to the receiving elevator to deliver the seed to the dryer or (2) by a separate elevating and conveying system. After drying is completed, the seed is returned to the raw seed storage facility to await conditioning. For such storage, some plants use bulk forklift toteboxes instead of separate bins. Equipment required include the truck access into the building, an under-the-floor receiving hopper of adequate size and design, conveyors, and forklifts to take seed dumped directly from trucks into the receiving hopper; conveying systems to carry seed to the desired storage bins (or forklift toteboxes); and a conveying system to carry the seed stored in receiving storage to the receiving hopper that then feeds the seed into the receiving elevator that carries the seed into the conditioning operations (equipment line).

7.E Bulk in forklift toteboxes

Some centers haul (by truck) empty forklift toteboxes to the contract grower's farm and directly fill the toteboxes on the farm, usually even filling the toteboxes while they are on the truck. The filled toteboxes are then trucked back to the conditioning center and handled by forklift in and out of raw seed storage.

7.F Receiving raw seed delivered in bags

In earlier times, the primary handling system for raw seed was to put the seed into bags on the farm and then truck the bags to the Seed Center, where forklifts unloaded the bagged seed and moved it into the raw seed storage building to await conditioning. This system is still used for lots/varieties that involve only small amounts of seed. It is expensive in that bags can seldom be used again, thus requiring a considerable numbers of bags.

7.G Receiving facilities

The receiving system must adequately physically receive all the different lots/varieties of seed and keep them completely identified and separate until it is moved into the conditioning line. In addition, complete records must be maintained to detail all operations conducted and condition of the seed lot.

7.H Ensuring identity

If identity (e.g., kind, variety, hybrid, lot) of seed is lost, the seed loses its value and must be discarded. Every container of seed (bag, bin, box, bulk)—any seed—must be completely labeled and must be completely identified in corresponding records that cover the seed.

7.H.1 Why is complete identity of seed so critical?

7.H.2 Why is a complete set of records, labels, and seed lot identification so essential? Describe.

7.H.3 How can the lot identification be labeled on each container of seed? Describe.

7.H.4 What should be done with a container of seed whose identity is lost or is uncertain?

7.I Sampling and quality testing

Adequate systems for receiving raw seed; inspecting it immediately upon arrival; and arranging immediate implementation of the required handling, labeling, drying, storage, and separation are critically essential. These systems should be developed along with implementation procedures and responsibilities in significant detail before the Seed Center becomes operational, so that all operations from day 1 ensure the complete identity, purity, quality, and condition of each lot and container unit of seed.

7.I.1 Describe a complete sampling and testing system and its operation.

7.J Drying

Generally, at least some seed drying must be done each season. In some cases, all seed received must be dried immediately upon arriving at the Seed Center.

The need for immediate and adequate seed drying is critical to seed quality and must be implemented. Quantity of seed and the extent of drying needs, number and size of lots to be dried, weather at time of drying, and drying systems most suitable must be described in advance. In addition, adequate facilities must be provided, with drying included in the operational program and schedules of the Seed Center. Safety requirements—for fire prevention, seed condition, and operations—must be planned and implemented effectively and completely. A complete system of checking seed moisture content at all stages of operation must be developed and implemented. The drying system is most commonly a bulk bin drying system. Seed is dumped into the drying system and properly dried. Seed may then be fed directly into conditioning, or it may be returned to raw seed storage to await conditioning.

7.J.1 List the types of dryer design and construction. Explain why a Seed Center may have several different drying systems.

7.J.2 Describe, and sketch the layout, of a receiving storage and drying for nonconditioned seed that is received in bulk or in bags or forklift toteboxes.

7.J.3 List in sequence the places/stages when seed moisture content should be checked.

7.K Nonconditioned seed storage

All seed comes into the Seed Center as nonconditioned seed that must usually be dried, stored, handled, and protected until it is conditioned. Experience has overwhelmingly shown that nonconditioned seed and conditioned seed should be kept in separate storages. Nonconditioned seed storage should be designed for efficient reception (unloading, moving into storage, and drying) and subsequence maintenance of all lots completely separate and pure.

Unprocessed seed should be stored in a separate place; it should never be stored with processed seed, because of the risk of bringing high moisture and insects to the already-conditioned seed.

Raw seed storage should be readily reached from the entrance into the Seed Center and handy for trucks. The raw seed storage should have two different delivery access systems: (1) one end of the storage (the end farthest from the seed conditioning machines) should adjoin the seed conditioning facility. At this end, there should be a large under-the-floor receiving pit serving the receiving elevator. Trucks bringing raw seed can drive over this dump and dump their seed into the pit. (2) Seed can also be brought or conveyed to this pit from the adjoining raw seed storage. This access system will allow a single receiving hopper and elevator to do double duty, as it can not only receive bulk raw seed delivered to the Seed Center in trucks and elevate it into bulk bins or toteboxes for storage, but also receive raw seed from the raw seed storage facility and elevate it into the seed conditioning line.

7.K.1 Sketch the layout of a receiving storage and its required equipment and facilities that can efficiently receive and handle raw seed delivered in bulk in trucks, and seed delivered in bags, toteboxes, or both.

7.K.2 *Detail why conditioned and unconditioned seed should not be stored together. Can you think of more reasons?*

7.K.3 *Describe the spouting, conveyors, and components of the elevator that receive raw seed and send it into bulk storage and that can also receive raw seed and spout it into the seed conditioning line.*

chapter eight

Moving raw seed into conditioning

8.A *Setting up machines to prepare for conditioning*

Before conditioning begins, each lot of seed should be sampled and given a good examination of the conditioning needed. This examination is necessary because each lot is different in the conditioning needed to remove undesirable materials and bring the lot up to the required quality standards.

And, if crop or variety or class of seed is different from that of the last lot handled, a detailed, complete clean-out of all machines is absolutely necessary.

8.A.1 *Why should each lot be sampled and analyzed to determine conditioning needs?*

8.A.2 *Why is a COMPLETE clean-out necessary?*

8.B *Sampling raw seed to determine conditioning needed*

Each lot of raw seed is sampled, before conditioning, and a complete Conditioning Needs Test is conducted on the sample, by using small-scale separators and the operator's knowledge of what each machine is capable of doing. This test tells what machines should be used and their position in the "conditioning line."

8.B.1 *Why is sampling and identifying the conditioning needed on every lot of seed?*

8.C *Checking flow sequence setup to ensure complete conditioning*

After determining the machines needed to condition the lot, examine the plant cleaning line to be sure that all needed machines will be used and that the seed will flow in and out of these machines properly.

8.C.1 List the conditioning machine sequence required for major crop seed in your area.

8.D Setup to handle conditioned seed as it is bagged

Carefully examine the end of the conditioning line to be sure that conditioned seed flow into the bagging setup and that space, handling, and facilities are set up to handle the bagged conditioned seed efficiently.

8.D.1 What has to be done to ensure that bagged conditioned seed is efficiently moved from the conditioning line into conditioned seed storage?

8.E Handling waste products

Examine each separating machine to be sure that the undesirable materials removed will flow efficiently into the plant's system for handling waste products.

8.E.1 Describe how each machine is examined to check that it has been properly cleaned out and set up to handle the new seed variety, and how waste products will be handled.

8.F Conditioning plan and schedule

To prepare the Seed Center's entire annual total of seed, a conditioning plan is usually prepared so that all seed can be received, handled, and conditioned in the most efficient and economical manner. For example, the first crop variety to be sold is conditioned first. All lots of this variety are conditioned in an unbroken sequence, so that it is not necessary to do a detailed clean-out of the conditioning plant after each lot prevents mixing or contaminating the seed with seed of another crop variety. Conditioning in this manner then requires only one clean-out after all seed of each variety is finished conditioning. This approach can save a huge amount of time and labor.

A detailed schedule of which seed lots are to be conditioned when, and when they must be ready for marketing, is essential. The schedule, as much as possible, plans for different lots of the same crop variety to be conditioned in an uninterrupted sequence to eliminate the need for a complete clean-out of the conditioning machines between lots.

Careful planning can save much time and expense. The major element in planning is to condition all lots of the same kind and variety in one continuous sequence. This approach eliminates the need for stopping and completely cleaning out the conditioning plant to avoid mixing varieties. The first seed variety to be marketed should also be the first seed variety conditioned.

8.G Moving raw seed into conditioning

According to the schedule, seed lots are moved from raw (not yet conditioned) seed storage into the conditioning area. The raw seed is moved to the receiving hopper that then feeds the nonconditioned seed into the receiving elevator. The seed conditioning "line" of machines must be fed continuously at the same rate (capacity) in a nonstop flow. If separating machines do not have the proper uniform and constant feed rate, it not only wastes time but also does not achieve the same separation.

So, raw seed must be moved from storage into the receiving elevator at a constant, nonstop, nonvariable feed rate.

The same feed hopper and elevator are normally used to receive incoming raw seed and move it into storage, and then also to feed raw seed into the conditioning line of machines.

To accomplish this goal, the raw seed storage is located adjoining the beginning of the seed conditioning machine layout.

8.G.1 *What are the most important elements in planning the sequence in which crop varieties will be conditioned?*

8.G.2 *Prepare a conditioning plan for a season when four different varieties of each of five different crops will be received, dried, and conditioned.*

8.G.3 *Describe how this elevator/receiving hopper/spouting system is set up. Sketch seed flow patterns.*

8.G.4 *How are raw seed receiving/drying, storage, and feeding seed into conditioning located? What raw seed receiving and handling factors are important in designing and operating the plant?*

chapter nine

Conditioning

9.A Seed conditioning sequences

Conditioning is the actual process of running a seed lot through a series (in a specific sequence) of separating machines, with each machine removing/separating a specific type of undesirable seed, materials, or both. Because of their different physical nature and different undesirable components, each crop seed has its own specific series of separating machines in the conditioning "line" or sequence of conditioning.

A single building is commonly used for raw seed storage, conditioning, and then conditioned seed storage. Operations move in this sequence receiving, drying, conditioning, and conditioned seed storage to maximize operating efficiency and minimize building, equipment, and handling costs.

9.A.1 *Give the specific series of separating machines used to separate undesirable materials from the five major crop seed planted in your area.*

9.B Handling and storing conditioned seed

When a seed lot has been conditioned, it should be pure, of high quality, and ready to be planted by the farmers who will use it. Before it is delivered to the dealers who serve these farmers, or before it is delivered directly to the farmers, it must be stored under carefully kept conditions so that it is not damaged, contaminated, or caused to lose germination. This is where the entire science of seed storage is focused.

9.B.1 *Describe the conditions desired for seed storage.*

9.C Storing and shipping conditioned seed

Conditioning is usually set up, and the plant is designed to expedite this conditioning, so that seed is handled in one continuous sequence of being received, dried as required, stored in the raw state (not conditioned), and then moved into the conditioning line for the required conditioning. Then, the seed is bagged and moved into conditioned seed storage until it is shipped out to the market.

After it is properly conditioned and bagged in the final bags that will be used to deliver the seed to retail dealers and using farmers, seed normally goes into conditioned seed storage to await transport to the markets. Seed is commonly kept in the conditioned seed storage until just before the planting season, because this is usually the best storage available, and also marketers often do not know in advance exactly how much of each kind of seed they need.

The conditioned seed storage usually has its floor raised to the level of the beds of trucks that are used to transport and deliver seed. This allows forklifts and workers to move seed directly from storage and go onto the truck to stack seed efficiently on the trucks. Conditioned seed storage is also usually nearest the gate of the Seed Center's compound to minimize the movement of trucks in the center and to simplify control and identification of trucks.

9.C.1 *Lay out a plant facility showing the unbroken sequence of operations of handling seed from receiving to shipping out to the market.*

9.C.2 *Describe a well-managed system for receiving orders for seed from retail dealers, identifying the seed to be sold, to each, then issuing written loading orders, loading the trucks, and clearing the loaded trucks for departure from the Seed Center.*

chapter ten

Support operations

10.A Plant area and wall

Control of mice, rats, birds, and insects is a major activity at a Seed Center. The entire (or at least most of) Seed Center's area should be covered by smooth-finished concrete that is poured and reinforced adequately for it to handle loads and to permit short-term seed handling activities without contaminating or adulterating the seed. This allows outside areas to be adequate for such operations as drying and temporary storage without damaging the quality of the seed.

It also makes it easy to clean up this area and eliminates cover and access for rats. An adequate concrete (or brick) wall around the Seed Center area that is joined to the concrete-paved inside area helps keep out rats, mice, and other pests and provides a safe operating area within the Seed Center's area.

10.A.1 *Describe some of the emergency or temporary seed handling, storage, loading, unloading, and drying that can be done in a properly paved and protected outside area within the Seed Center.*

10.B Truck scales to weigh incoming and outgoing trucks

Immediately after the entrance gate and driveway into the Seed Center, the truck scale should be installed so that all incoming trucks bringing raw seed can be conveniently weighed as soon as they arrive, and loads of outgoing conditioned seed can be weighed as they leave the Seed Center.

The truck scales should be installed beside the entrance driveway, so that vehicles that are not to be weighed do not have to pass over the truck scale. Scales should also be located beside the main office building, so that office staff can serve double-duty and operate the scales and truck weighing system.

10.B.1 *Sketch the layout of the Seed Center, including the truck scale and route to the raw seed storage.*

10.C Management and administrative offices

The office facility is usually located near the entrance into the Seed Center compound. The truck scale is located beside the office building, so that the Seed Center's secretary or bookkeeper can do double-duty by operating the truck scale and recording weights.

10.C.1 Why is the truck scale located beside the main office building?

10.D Internal quality control lab, staff, and procedures

The internal quality control (IQC) program is a critically important element. It works with all Seed Center personnel and operations to ensure that final seed quality meets or exceeds the required standards and that operations do not result in excessive loss of good seed. Every operation and machine adjustment is checked by IQC at frequent and regular intervals, and any necessary changes are identified and discussed with the concerned staff. Each fraction of good seed and waste materials from each machine or operation is checked for good seed purity and quality, and waste products are checked for amounts of good seed lost. Many samples are drawn each day, and checked or tested for purity, germination, and moisture content.

To ensure high quality of all seed, the conditioning plant must have an IQC section with adequate staff competent to conduct the necessary inspecting, sampling, testing, and recording at all stages of operations from incoming raw seed to conditioning, handling, storing, delivering, and all other operations.

Plant IQC staff should include at least two trained and competent seed analysts, plus two to four trained subprofessional staff who are capable of conducting the operations of sampling, purity testing, and germination testing. In off seasons when IQC work is low, these staff members usually are "multiduty" and conduct activities such as working with contract growers, helping in seed marketing, and promoting farmer acceptance and use of improved seed.

IQC must have a testing laboratory with complete equipment and facilities for sampling bulk and bagged seed and conducting tests for purity, noxious weeds, germination, seed treatment coverage, insect damage, and vigor. Some equipment needs are listed later in this manual.

The IQC lab is usually located with the main offices. However, the analysts must have ready and constant access to raw seed storage, conditioning, drying, and conditioned seed storage.

The IQC must have a fully equipped quality testing laboratory. IQC test results should carry the same level of correctness and acceptance as the external testing laboratory in the area. Because equipment is usually small and can be installed in space similar to offices, the IQC lab is usually installed in the same building as the main office.

10.D.1 List the primary operations of the IQC section. Where and why does it sample seed? What QC tests must the IQC be able to conduct?

10.D.2 What QC equipment and sampling/testing skills does the IQC need?

10.D.3 Describe the role of IQC in seed conditioning.

10.D.4 List the sequence (and purposes) of sampling each seed lot from first introducing it into conditioning until the seed is completely conditioned and filled into the bags in which they will be marketed.

10.E Truck parking, loading, inspecting loads, servicing, and handling

Raw seed is transported by trucks from growers to the Seed Center, and conditioned seed is transported by trucks from the Seed Center to the wholesale and retail dealers and to the farmers who will use the seed to plant their crops. The Seed Center must be set up and operated so that trucks and their loads can be handled efficiently. For example, raw seed brought in bulk requires the raw seed receiving/storage building to have an in-the-floor inside the building receiving hopper setup so that trucks can drive over the top of the receiving hopper and dump their loads. To receive raw seed brought in bags and/or totelift boxes, the raw seed storage should have its floor level at the height of truck beds, so forklifts can readily go onto flat-bed trucks to handle seed loads.

There must be facilities within the Seed Center to handle trucks; load and unload them; and keep trucks safely in covered parking sheds during idle seasons to hold loaded trucks safely before they are unloaded or shipped out and to maintain and repair trucks during off-seasons.

10.E.1 Describe the role of trucks in the seed conditioning operations.

10.F Workshop

The Seed Center uses machines and trucks in its operations. To be efficient and economical, the Seed Center must have the tools and facilities required to handle at least the basic maintenance and repairs on seed machinery and trucks. Most Seed Centers also have a workshop with the capability to build its own forklift pallets and totelift boxes, and they have the basic tools required to prepare and maintain machines, bins, and installation and operational facilities.

10.F.1 Describe the types of work, construction, and maintenance that can be done by Seed Center staff to maintain operations at minimum operating costs.

chapter eleven

Seed conditioning principles

11.A *Basic concepts*

11.A.1 *What is seed conditioning?*

11.A.2 *Why are seed conditioned? Cite at least six reasons.*

11.A.3 *How are undesirable materials (seed or particles) separated from good seed?*

11.A.4 *List and describe nine physical differences or characteristics used to separate seed mechanically.*

11.A.5 *Why are seed usually conditioned over a series of machines, instead of only one machine?*

11.A. 6 *Illustrate, with simple diagrams, each physical characteristic used to separate seed. List the machines that use each characteristic to separate seed.*

11.A.7 *Diagram the complete flow sequence of seed conditioning.*

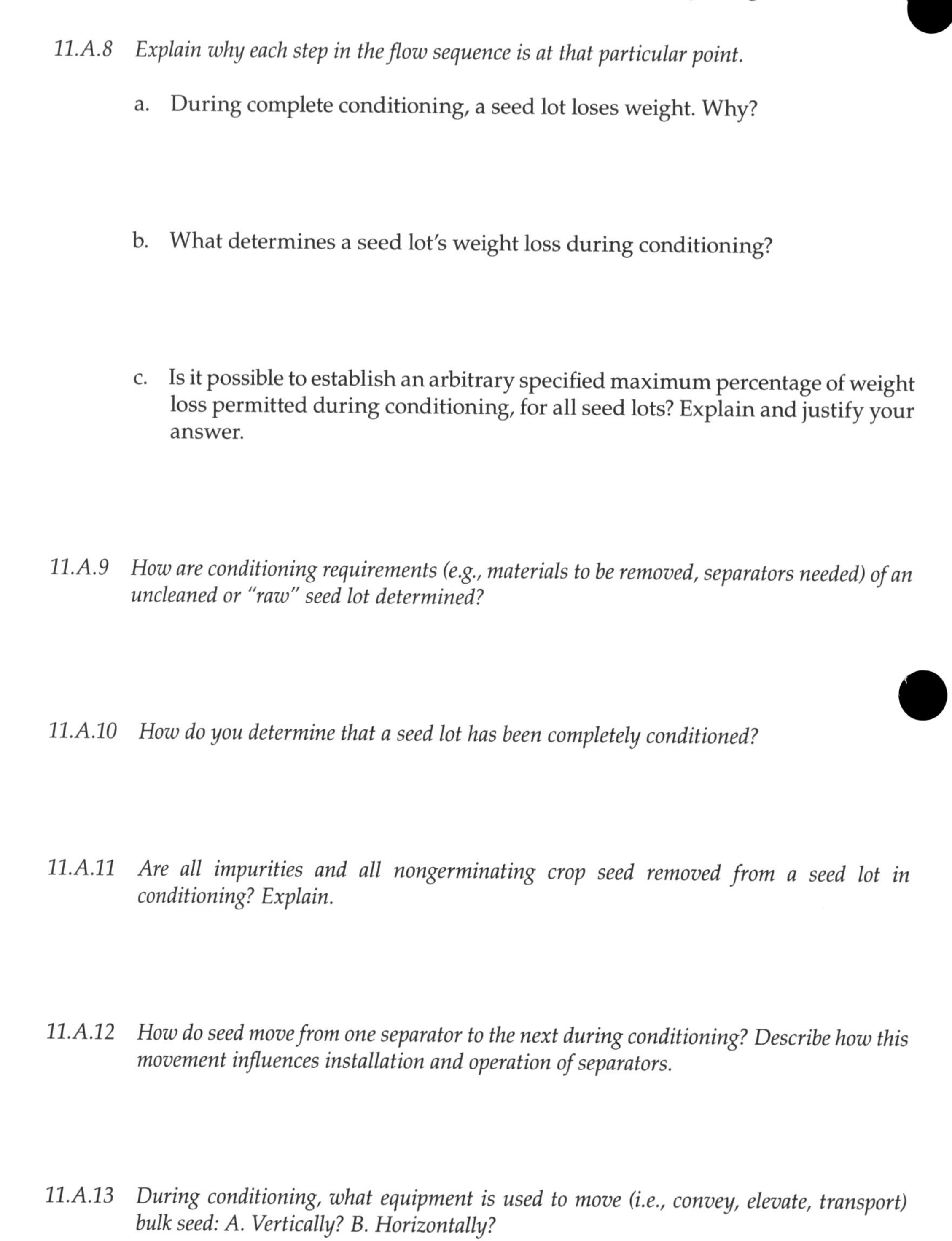

11.A.8 Explain why each step in the flow sequence is at that particular point.

a. During complete conditioning, a seed lot loses weight. Why?

b. What determines a seed lot's weight loss during conditioning?

c. Is it possible to establish an arbitrary specified maximum percentage of weight loss permitted during conditioning, for all seed lots? Explain and justify your answer.

11.A.9 How are conditioning requirements (e.g., materials to be removed, separators needed) of an uncleaned or "raw" seed lot determined?

11.A.10 How do you determine that a seed lot has been completely conditioned?

11.A.11 Are all impurities and all nongerminating crop seed removed from a seed lot in conditioning? Explain.

11.A.12 How do seed move from one separator to the next during conditioning? Describe how this movement influences installation and operation of separators.

11.A.13 During conditioning, what equipment is used to move (i.e., convey, elevate, transport) bulk seed: A. Vertically? B. Horizontally?

11.A.14 *Why is a holding bin normally used above each separator?*

11.A.15 *In general terms, how is the required capacity of a holding bin determined?*

11.A.16 *What is the actual capacity of a holding bin? How do you determine it?*

11.A.17 *Why are seed treated? What kinds of treating materials may be applied?*

11.A.18 *Why are cleaned seed usually bagged when conditioning is completed?*

11.A.19 *What kinds of packages are used for seed? How are the kind and size of package determined?*

11.A.20 *To understand and operate a seed conditioning machine properly, the operator must know three things:*

a.

b.

c.

11.A.21 *How are undesirable or waste materials that are separated from the crop seed handled, conveyed, held under separate conditions, and identified?*

11.A.22 *Other points, conditions, or elements of special interest during operations.*

chapter twelve

Waste products

12.A Waste product creation

Conditioning's objectives are to remove undesirable materials from the seed, so that the final product is pure, high-quality seed. In the conditioning operations, different separating machines are used to remove specific undesirable seed and materials from a lot of seed. This process creates a fraction of material termed 'waste products' that are not sold as seed. This material may be used as livestock feed, or it may be disposed of by burning or by dumping in a waste disposal area.

If seed has been harvested and handled carefully, waste products are usually not more than 5%–10% of a seed lot. However, with some crops such as chaffy grass seed, 50% or more of the raw seed lot will be removed as waste products.

Handling separated waste products must be planned and implemented carefully to get the waste materials out of the way quickly and completely and to minimise risks of wastes being a potential contaminant. Some plants return waste products to the contract growers, but the most efficient method is to have a contract with a livestock feed supplier who can remove all the waste products and blend them into the livestock feeds for ultimate disposal. At the Seed Center, waste products are not stored in the conditioned seed storage; wastes are generally kept in a separate storage, but they may also be kept in the raw seed storage.

12.A.1 Describe a well-managed system for handling and disposing of separated waste products.

chapter thirteen

Sanitation and pest/insect control

13.A Cleanliness

All seed facilities must be kept clean and orderly to both minimize cover and attraction for rats, mice, insects, and birds and prevent accidental contamination of the seed lot being handled.

In addition, exclusion and trapping programs for rats and birds much be maintained, and spraying programs for insect control must be constantly ongoing.

13.A.1 *Describe the organization and implementation of a complete program to minimize damages from insects, rats, and birds.*

13.B Sanitation and pest/insect control

The Seed Center must be kept clean to prevent undesirable materials from being accidentally mixed with the crop seed.

Cleanliness also is required to help minimize attraction for rats and birds and to help control insects.

A definite ongoing sanitation and insect and rat/bird control system must be carefully and completely maintained.

13.B.1 *Design a plan for control of insects, birds, and rats for a Seed Center. Include sanitation, exclusion, poison baits, poison control, and protection of personnel.*

chapter fourteen

Conditioning equipment layout

14.A Equipment layout

Layout can be defined as the sequence of conditioning operations and equipment installed in the seed conditioning plant. Layout varies widely not only because of the volume of operations but also because of the different crop seed handled: different crop seed require different separations that, in turn, require different separating machines.

A seed conditioning plant used to remove undesirable seed and particles to improve the purity and quality of crop seed uses a specific sequence of separating machines to make a series of specific separations. The sequence of separations used for each crop seed has been developed through experiences over the years in improving crop seed quality. This sequence has involved identifying the specific contaminant materials usually associated with each crop seed, determining the usable physical differences between the crop seed and the undesirable seed/particles that must be removed, and then using the appropriate seed separating machines installed in the sequence that permits each machine to perform best. Years of seed conditioning experience have resulted in developing specific systems for efficient conditioning of each crop seed. As shown in Figure 14.1, the separators needed, and the most efficient sequence of cleaning operations, usually includes the following:

1. A receiving pit and elevator, and other equipment such as conveyors, bins, and spouting, that receive the seed to be conditioned and introduce it into the series of machines that separate undesirable seed or particles so that the final conditioned seed lot contains only the desired crop seed, of high quality and purity.
2. The air-screen cleaner is normally the machine (separator!) that does the basic cleaning by using a series of screens with perforations or openings, ranging from one or two screens in a small scalper-cleaner up to 10 or 12 in a high-capacity cleaner, plus one, two, or three air blasts to remove lightweight (i.e., low-specific gravity) undesirable seed or particles.
3. One, two, or more additional separators that make further separations by using a specific physical difference in seed to separate and remove specific contaminants that possess a specific physical difference from the good crop seed. For example, a length separator is used for cleaning small-grain seed (wheat, barley, oats, rice) because the most common contaminants are seed that possess different lengths; a gravity separator is included for making a gravity separation when separation of lighter or heavier undesirable seed is needed; a roll mill is used for small legume seed because rough-surfaced dodder seed is a common contaminant; and a color sorter is used for seed such as large beans or peas because undesirable contaminant seed is often of different colors.
4. The seed conditioning layout commonly includes a seed treater to apply a protective chemical to the seed.
5. The final machine/operation is usually a bagging or packaging system that packages the seed in desired quantities in protective containers.
6. Other machines can be added as needed, in appropriate positions in the "cleaning line."

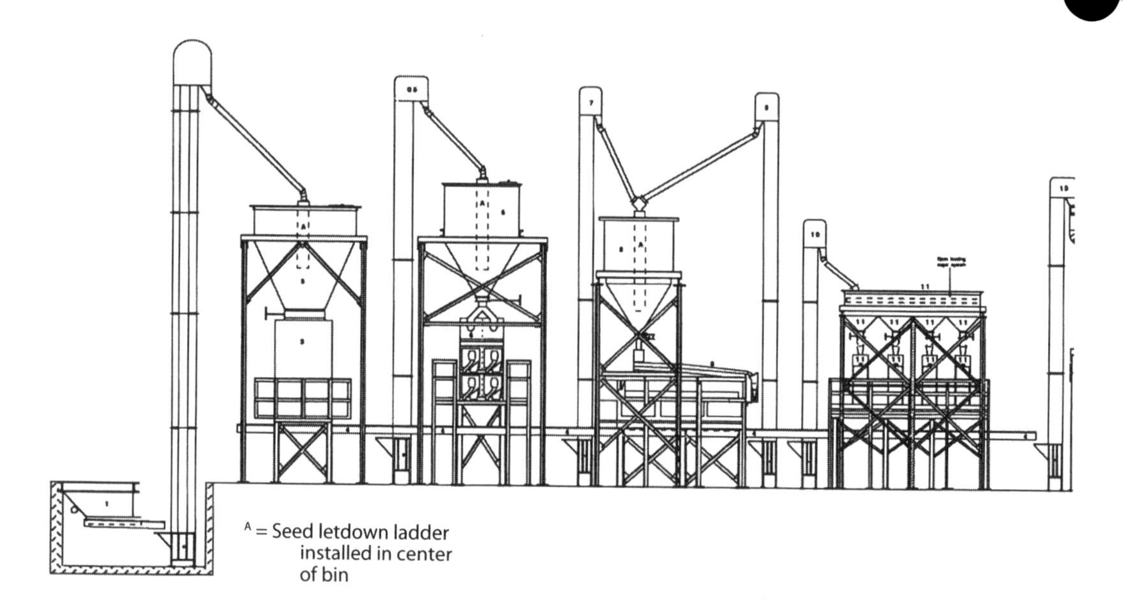

Figure 14.1 Basic layout sequence of a seed conditioning plant. Starting on the left side, uncleaned seed is fed into the receiving hopper of the receiving elevator. This elevator delivers the seed into the holding bin over the air-screen cleaner, the basic cleaner that is usually the first separator. A separate holding bin is used to permit feeding each separator at its most efficient rate; an elevator (often combined with suitable conveyors and facilities to prevent mechanical damage to seed) carries seed from each separator to the next separator or machine in the flow sequence (in this layout, the roll mill used to remove dodder seed from alfalfa seed); an elevator carries the seed to the gravity separator. Then, an elevator carries the seed to the bins that accumulate the cleaned seed before sending them to the elevator that feeds bagging/weighing (not shown here).

14.A.1 *Why is a sequence of separators used and such an important part of seed conditioning?*

14.A.2 *Other points, conditions, or elements of special interest during operations.*

chapter fifteen

Receiving pit

15.A *Receiving pit*

Procedure: Study the receiving bucket elevator(s) and answer the following questions.

15.A.1 *What is the purpose of the receiving pit and its associated components?*

15.A.2 *Where and how is the receiving pit installed? List and describe the other equipment, components, and items involved in making the receiving pit adequate for efficient receiving of unconditioned seed and delivering it into the conditioning "line."*

15.A.3 *How is the required capacity of the receiving pit determined, and where seed is received (1) in bags or forklift toteboxes, or (2) in bulk in trucks?*

15.A.4 *What is the most efficient shape of the receiving pit? Why or what makes this an efficient shape?*

15.A.5 *How is the receiving pit made and located so that trucks can dump directly into it?*

15.A.6 *How is seed emptied from the receiving pit? How is this designed and located so that the pit is completely self-emptying?*

15.A.7 *What is seed from the pit fed into? What other equipment is required and installed with the pit to deliver seed into the conditioning operation? How and where is this equipment installed?*

15.A.8 *What is the sequence of machines and spouting that handle different crop seed as it moves from the receiving pit to the air-screen cleaner?*

15.A.9 *Why is it essential to have easily accessible worker access to the outside bottom of the receiving pit?*

15.A.10 *Other points, conditions, or elements of special interest.*

15.B *Receiving installation*

Because receiving is the first instance requiring intimate knowledge of the installation, part of the answer to this question on the critically important installation is included here.

1. A pit of adequate size (capacity) must be excavated in the floor of the plant's building. The pit should be in a location that would allow trucks bringing in raw seed to drive into the building, pass over the receiving pit installed in this floor pit, and stop in a position that will allow the unconditioned seed to be dumped from the truck into the pit installed below the floor level. The walls and floor of this concrete pit are made of reinforced concrete construction and are smooth finished.
2. The actual seed receiving pit, which receives and handles the seed, is made of steel with smooth internal surfaces so that no seed will be held by uneven places in the receiving pit walls. This pit extends up to the floor level, to the top of the newly constructed concrete pit in the floor of the building.
3. The top part of the receiving pit is at least 8 feet wide and 8 feet long in the direction of the truck's movement. The receiving pit is like a bin in the floor. The top part of the seed-holding receiving bin (NOT the concrete pit constructed to hold the receiving bin) has straight-down walls that extend as far as necessary to give the pit the

required holding capacity. Note: In this example, the pit's top part is square, with dimensions of 8 feet on all four sides. This square construction extends downward for 4 feet, where the pit construction begins to slope into a manageable size at the bottom discharge point. The top square part, combined with the sloping bottom of the pit, will hold the required amount (one truckload) of seed if the straight side walls are 4 feet from top to bottom.

4. The end walls of the seed receiving pit (on the end next to the conditioning plant equipment and the end away from the conditioning plant equipment) are not sloped, but they are cut to give the assembled receiving pit the shape of a giant wedge. These walls (cut so this lower part of the pit extends inward) go straight down to the bottom end of the seed pit. The two sides of the seed pit that are located on the sides over which the truck passes are sloped inward at 60° so that the bin resembles a huge "wedge." The bottom of this wedge-shaped bin is 1 foot wide and is open the entire width of the receiving bin.
5. No shutoff gate is placed on this open bottom of the receiving pit. A vibrating conveyor with a vibrating seed movement channel (not less than 14 inches wide) is installed below this open base of the pit, and it forms the bottom of the seed receiving pit.
6. The vibrating conveyor forms the bottom of the receiving pit. It extends across the entire bottom of the receiving pit and extends outward from one side of the pit so that it extends outward to the intake hopper of the receiving elevator that takes seed into the conditioning line. When this conveyor is turned on, it takes seed from the receiving pit, transports it, and discharges it into the receiving hopper of the receiving elevator. The elevator then takes the seed to the desired machine to initiate the desired conditioning.

   Note: This vibrating conveyor is turned on only after the receiving elevator is turned on and the plant is set up to receive, handle, and condition the seed.

7. An up-and-down sliding shut-off gate is installed above the vibrating conveyor at the side end of the receiving pit so that the flow of seed into the receiving elevator is prevented.
8. A grille of welded round steel pipes with no surfaces that could catch and hold seed and that is strong enough to support the weight of the loaded truck is installed at floor level, so the truck can safely drive over the pit and dump its load of seed.
9. The operating sequence is as follows: the discharge spouting of the receiving elevator is set to send the incoming seed to the desired place; the receiving elevator is turned on; and the vibrating conveyor is turned on, so that it can take seed from the receiving pit and deliver it into the receiving elevator. The receiving elevator's discharge spouting is set so that seed is delivered to the desired first machine in the conditioning line. See receiving elevator discharge spouting below.
10. A smaller channel or pit is constructed below the floor level and extends from the receiving pit to the pit in which the plant's receiving elevator is installed. This pit contains the part of the vibrating conveyor that extends from the receiving pit to the plant's receiving elevator, and it allows seed to be delivered straight from the receiving pit to the receiving elevator.
11. The concrete pit must have adequate space for workers to enter and clean out the bin and conveyor as required. The vibrating conveyor is mounted on a frame that holds it up to the bottom of the receiving bin. Electric lights and plug-in outlets must be provided in the concrete pit to facilitate workers in this area and to provide electricity for equipment such as a vacuum cleaner.

12. Discharge spouting on the receiving elevator is set according to the kind of crop seed being conditioned and its needs for conditioning as follows:
    a. A two-way valve (no. 1) is attached to the discharge spout of the elevator.
    b. Additional two-way valves (no. 2 and no. 3) are attached to each of the discharge spouts of the first two-way valve.
    c. Spouting attached to the two outlets of two-way valve no. 2 can send seed to either the debearder or the huller-scarifier. Oat, barley, and chaffy grass seed is sent to the debearder through outlet 1 of two-way valve no. 2; some small-seeded legumes (e.g., some clovers, alfalfa) are sent to the huller-scarifier through outlet no. 2 of two-way valve no. 2.
    d. Spouting attached to the no. 1 outlet of two-way valve no. 3 can send seed either to the scalper or directly to the air-screen cleaner.
    e. To summarize:
       Spout 1 of two-way valve no. 2 feeds oat, barley, and chaffy grass seed to the debearder.
       Spout 2 of two-way valve no. 2 feeds small legume seed to the huller-scarifier.
       Spout 1 of two-way valve no. 3 feeds any seed that has much plant material to be removed to the scalper.
       Spout 2 of two-way valve no. 3 feeds "ordinary cleaning required" seed to the air-screen cleaner.

15.B.1 *How is the required capacity of the receiving pit determined, and where seed is received (1) in bags or forklift toteboxes, or (2) in bulk in trucks?*

15.B.2 *What is the most efficient shape of the receiving pit? Why or what makes this shape efficient?*

15.B.3 *How is the receiving pit made and located so that trucks can dump directly into it?*

15.B.4 *How is seed emptied from the receiving pit? How is this designed and located so that the pit is completely self-emptying?*

15.B.5 *What is seed from the pit fed into? What other equipment is required and installed with the pit to deliver seed into the conditioning operation?*

chapter sixteen

Bucket elevator characteristics

Conditioner name:

Date:

Seed conditioning plant/machine location:

Machine brand, model number, other identification:

Year machine installed; condition:

Installation and feeding seed/waste materials to/from the machine description:

Any problems encountered with operation of the machine:

Procedure: Study the bucket elevator(s) and provide the following information or answers to questions.

1. Make and model.

2. Name and address of manufacturer.

3. Is an operator's manual available? What is the machine designed to do? Is it being used for its designed purpose? If not, how is it being used? Can its use be changed to take better advantage of its design features? Is another machine(s) needed to improve cleaning performance?

4. Dimensions:

   a. Overall height.

   b. Discharge height (total seed elevating height from elevator base to discharge spout).

   c. Boot height (base to top of seed intake hopper).

d. Effective seed elevating height (of seed intake hopper to discharge spout).

e. Head space (from discharge spout to top of elevator).

f. Leg: width and length.

g. Head and discharge spout width and length.

5. How is the elevator mounted and supported?

6. How is access into the elevator boot provided for clean-out and other activities? Describe.

7. a. Boot belt pulley dimensions:

   - Diameter

   - Width

   - Shaft diameter

b. Describe boot belt pulley type, and its clean-out characteristics.

8. Draw the elevator hopper that feeds seed into the elevator. Show all dimensions. What is the capacity of the hopper?

9. How is rate of feeding seed into the elevator controlled?

10. How much space is between the buckets and the bottom of the boot? Describe.

11. a. Is let of the boot open or enclosed? Is each leg enclosed separately? Describe.

    b. Are inspection doors provided in the leg? Describe.

12. Describe and give dimensions of the belt carrying the buckets.

13. Buckets (cups):

- Width
- Height

- Depth
- Center-to-center spacing
- Shape
- How are buckets attached to the belt?
- Are spacers used between the buckets and the belt? On the bolts? Why? Describe.
- Draw a bucket, to illustrate its shape.

14. Describe access to the elevator head. Is it readily and safely accessible? Does it adequately provide access to clean out and inspect the head?

15. Power required.

16. Diagram the motor and power delivery system. Show diameter and RPM of each pulley.

17. Motor RPM. Elevator head pulley RPM.

18. How are motor and drive belts tightened? Describe.

19. Head pulley

    Diameter

    Width

    Shaft diameter

    RPM

    Calculate feet per minute of belt speed

    Type and construction of head pulley

20. Can the elevator head cover be removed easily? Are inspection doors in the head? Describe.

21. What kind of discharge spout is used?

22. Is an inspection door provided in the discharge spout? Describe.

23. Draw the head and discharge spout, showing the shape of the discharge side of the head.

24. How is the belt carrying the buckets tightened? Why is it sometimes tightened?

25. How are ends of the belt fastened together (to make an "endless loop")?

26. Is a dust exhaust connection installed on the elevator? Where? Describe.

27. Briefly discuss and describe other makes, models, heights, and types of bucket elevators available.

28. Points, conditions, or elements of special interest.

29. In the notebook that carries your study manual and your own observations, attach a copy of the operation and maintenance manuals of this model of machine that are installed in your facility.

30. Identify and describe five factors or conditions that influence the performance of this machine.

31. Where is seed handled by this elevator coming from, and where is it being fed to?

32. What additional facilities, structures, and subsidiary equipment are required for this machine to operate effectively and efficiently?

chapter seventeen

Bucket elevator operation

Conditioner name:

Date:

Seed conditioning plant/machine location:

Machine brand, model number, other identification:

Year machine installed; condition:

Installation and feeding seed/waste materials to/from the machine description:

Any problems encountered with operation of the machine:

Procedure: Using a seed lot, set up, adjust, and operate the bucket elevator. Provide the following information or answers to questions.

1. Seed crop and variety.

   Original seed lot:

   Weight

   % Mechanical seed damage

   % Broken seed

   % Damaged but unbroken seed (cut, etc.)

2. Elevator make and model.

3. Is a maintenance manual available, or an operator's manual with a detailed description of recommended maintenance? Is the machine in good condition and well maintained? What changes/improvements in maintenance would improve it?

4. How are seed fed into the elevator?

5. How are seed handled as they discharge from the elevator? How far do they fall? Do seed directly strike a metal surface at any point?

6. Discharge height of the elevator.

7. *Using extreme caution to ensure safety,* remove the head cover and observe as seed is discharged from the buckets.

   a. Do any seed "back leg," that is, fall back down inside the "rising" elevator leg? Describe.

   b. When the head cover is off, are seed discharged directly into the discharge spout, or do some seed fall outside? Describe.

   c. Does the discharge path of seed, and the appearance of the inside of the head cover, indicate that seed discharging from the buckets strike the head cover? Does this cause mechanical injury to seed? Describe.

8. Determine the time required to elevate a weighed quantity of seed, and calculate capacity per hour.

9. Make significant changes in adjustments as indicated below, and note the results. (After an adjustment is noted, return it to normal operation before changing the next adjustment.)

   a. Open elevator feed gate (in the feed hopper) to the maximum.

b. Slacken the belt by loosening belt-tightening bolts on the boot pulley shaft.

c. Start the elevator with the boot already filled with seed.

d. Plug discharge spout (e.g., as occurs when the fed bin fills up).

e. Loosen motor belt, or any drive belt.

10. Using head pulley RPM and diameter, calculate feet per minute of belt speed.

11. Open leg inspection plate (*caution: keep hands out*). Can seed and moving buckets be adequately observed? Are the buckets full? Do seed fall out of the buckets? Describe.

12. After all seed is elevated, stop the elevator and set the safety switch, so the elevator cannot be turned on.

    a. Open all boot inspection doors. Collect and weigh all loose seed remaining in the elevator.

b. By hand, turn the belt pulley. Carefully check 15 consecutive buckets to see whether any seed is lodged on, in, or behind the buckets. Discuss your findings.

c. Examine the boot pulley. Does it hold seed that may contaminate the next seed lot? How can it be cleaned out?

13. Points, conditions, or elements of special interest.

chapter eighteen

Vibrating conveyor characteristics

Conditioner name:

Date:

Seed conditioning plant/machine location:

Machine brand, model number, other identification:

Year machine installed; condition:

Installation and feeding seed/waste materials to/from the machine description:

Any problems encountered with operation of the machine:

Procedure: Study the vibrating conveyor and provide the following information or answers to questions.

1. Make and model.

2. Name and address of manufacturer.

3. Is an operator's manual available? What is the machine designed to do? Is it being used for its designed purpose? If not, how is it being used? Can its use be changed to take better advantage of its design features? Is another machine(s) needed to improve cleaning performance?

4. How does this machine convey seed? Describe the conveying action, slope, and design of the conveying mechanism.

5. Overall dimensions:

   Length

   Length of seed conveying

   Height

   Width

   Height: Floor to seed intake. Floor to seed discharge.

6. What are this vibrating conveyor's requirements for

   a. Power?

   b. Installation and mounting?

   c. Seed feed into the conveyor?

   d. Handling seed as they discharge from the conveyor?

7. Does the conveyor use an electromagnetic system, eccentric shaft, or other drive system? Describe its design and operation characteristics.

8. How is the vibrating motion created, to convey seed? Describe.

9. What is the maximum length of horizontal backward-and-forward vibrating motion of the conveyor?

10. Does the conveying trough/tube move vertically as it vibrates? Describe.

11. Is the seed conveying trough/tube open or closed on top? Describe, with advantages and disadvantages.

12. Draw a cross section of the conveying trough/tube. Show all dimensions.

13. Is the conveying trough/tube a single unit, or several sections bolted together? Describe.

14. Is this conveyor designed so that it can readily be shortened or lengthened? How could a shorter or longer conveying distance be achieved?

15. Is the tube/trough of the conveyor sloped (i.e., inclined from the feed intake to outlet)? How much? How is the slope created?

16. Can this conveyor move seed only horizontally, or can it also move seed up an incline? If so, what is the maximum angle of inclination at which it can convey seed?

17. Does the conveyor have adjustments to regulate its action? Describe.

18. Can more than one discharge point be served by this conveyor? Describe.

19. Briefly discuss and describe other sizes, models, and makes of vibrating conveyors.

20. Points, conditions, or elements of special interest.

21. In the notebook that carries your study manual and your own observations, attach a copy of the operation and maintenance manuals of this model of machine that is installed in your facility.

22. Identify and describe five factors or conditions that influence the performance of this machine.

chapter nineteen

Vibrating conveyor operation

Conditioner name:

Date:

Seed conditioning plant/machine location:

Machine brand, model number, other identification:

Year machine installed; condition:

Installation and feeding seed/waste materials to/from the machine description:

Any problems encountered with operation of the machine:

Procedure: Using a seed lot, set up and operate the vibrating conveyor. Provide the following information or answers to questions.

1. a. Seed crop and variety.

   b. Seed lot weight.

2. Conveyor make and model.

3. Is a maintenance manual available, or an operator's manual with a detailed description of recommended maintenance? Is the machine in good condition and well maintained? What changes/improvements in maintenance would improve it?

4. Conveying distance.

5. How are seed

   a. Fed into the conveyor?

   b. Handled as they discharge from the conveyor?

6. Is seed dust generated by the conveyor? How? How can it be controlled?

7. Do seed fall or bounce out of the conveyor as it conveys seed? Describe.

8. a. What is the rated capacity of the conveyor, per hour?

   b. What capacity was reached during this operation?

   c. Do you think the rated capacity is accurate? Explain.

9. Does the conveying action cause mechanical injury to seed? Explain.

10. Can seed be mechanically injured as they are fed into, or discharged from, the conveyor? What should be done to prevent injury?

11. Do seed lodge in the conveyor, so that the next seed lot may be contaminated? How is the conveyor cleaned out?

12. What are the uses and limitations of this conveyor in seed conditioning? Points, conditions, or elements of special interest.

chapter twenty

Horizontal belt conveyor characteristics

Conditioner name:

Date:

Seed conditioning plant/machine location:

Machine brand, model number, other identification:

Year machine installed; condition:

Installation and feeding seed/waste materials to/from the machine description:

Any problems encountered with operation of the machine:

Procedure: Study the conveyor and provide the following information or answers to questions.

1. Make and model.

2. Name and address of manufacturer.

3. Is an operator's manual available? What is the machine designed to do? Is it being used for its designed purpose? If not, how is it being used? Can its use be changed to take better advantage of its design features? Is another machine(s) needed to improve cleaning performance?

4. Dimensions:

   Overall height, width, length.

   Height from floor to seed feed intake, and to seed discharge.

   Distance seed is conveyed.

5. Describe the method or principle by which seed is conveyed.

6. What are the conveyor's requirements for

   a. Power?

   b. Installation and mounting?

   c. Seed supply and feed?

   d. Handling discharges seed?

7. Type of conveying belt.

   Belt width and thickness.

8. Is the belt troughed or flat? How are seed kept on the belt?

9. Is the belt smooth? Does it have conveying ribs? Describe.

10. Does this conveyor have idler pulleys? How is the belt supported?

11. How is the conveying belt tightened?

12. a. Diagram the power delivery system from motor to conveying belt drive pulley. Show diameter and RPM of each pulley.

    b. What is belt drive pulley RPM?

    c. What is speed of the belt?

13. Can seed fall under the conveying belt? How are, or can, they be removed?

14. Describe the conveyor's hopper to receive seed. Can multiple feed intakes be used? Do seed spill out?

15. Describe the conveyor's seed discharge.

16. Is the conveyor portable? Describe.

17. Briefly discuss and describe other makes, models, and kinds of horizontal belt conveyors available.

18. In the notebook that carries your study manual and your own observations, attach a copy of the operation and maintenance manuals of this model of machine that is installed in your facility.

19. Identify and describe five factors or conditions that influence the performance of this machine.

chapter twenty-one

Horizontal belt conveyor operation

Conditioner name:

Date:

Seed conditioning plant/machine location:

Machine brand, model number, other identification:

Year machine installed; condition:

Installation and feeding seed/waste materials to/from the machine description:

Any problems encountered with operation of the machine:

Procedure: Using a seed lot, set up and operate the conveyor as described. Provide the following information or answers to questions.

1. Seed crop and variety.

2. Conveyor make, model, and manufacturer.

3. Is a maintenance manual available, or an operator's manual with a detailed description of recommended maintenance? Is the machine in good condition and well maintained? What changes/improvements in maintenance would improve it?

4. Distance seed is conveyed.

5. How are seed

   a. Fed onto the conveyor?

   b. Handled when they discharge from the conveyor?

6. Is seed mechanically inured as it goes onto or from the conveyor? How can this be prevented?

7. Does the conveying process cause mechanical injury to seed? How? How could this be eliminated?

8. Do seed fall out of the conveyor? How much? How can this be prevented?

9. Do seed lodge in the conveyor, under the conveying belt, etc., so that the next seed lot may be contaminated? How is the conveyor cleaned out?

10. Does this conveyor move seed only horizontally, or also up an incline? If so, what is its maximum inclination?

11. What are the uses and limitations of this conveyor in seed conditioning?

chapter twenty-two

Inclined belt conveyor characteristics

Conditioner name:

Date:

Seed conditioning plant/machine location:

Machine brand, model number, other identification:

Year machine installed; condition

Installation and feeding seed/waste materials to/from the machine description:

Any problems encountered with operation of the machine:

Procedure: Study the conveyor and provide the following information or answers to questions.

1. Make and model.

2. Name and address of manufacturer.

3. Is an operator's manual available? What is the machine designed to do? Is it being used for its designed purpose? If not, how is it being used? Can its use be changed to take better advantage of its design features? Is another machine(s) needed to improve cleaning performance?

4. Overall dimensions:

   Width

   Length

   Height

   Conveying trough dimensions:

   Type

   Width

Length

Depth

Seed conveying distance

5. How are seed held in place on the belt as it travels up an incline? Describe.

6. Conveying belt:

Type

Width

Thickness

Height, spacing, construction of ribs, etc.

7. What are the conveyor's requirements for?

a. Power (including RPM)?

b. Installation and mounting?

c. Seed supply and feed?

d. Handling discharged seed?

8. a. Describe the drive system.

b. Draw the drive system. Show diameter and RPM of all pulleys.

9. Describe the pulleys on the upper and lower ends of the conveying belt.

10. How is the conveying belt tightened?

11. Are idler pulleys provided? How is the belt supported?

12. Is the conveyor portable? If so, how is it moved?

13. Can the belt move in only one direction, or in either direction? Describe.

14. Can seed get under the conveying belt? What provisions are made to remove seed that gets under the belt?

15. Briefly discuss and describe other makes, models, and kinds of inclined belt conveyors available.

16. Points, conditions, or elements of special interest.

17. In the notebook that carries your study manual and your own observations, attach a copy of the operation and maintenance manuals of this model of machine that is installed in your facility.

18. Identify and describe five factors or conditions that influence the performance of this machine.

chapter twenty-three

Inclined belt conveyor operation

Conditioner name:

Date:

Seed conditioning plant/machine location:

Machine brand, model number, other identification:

Year machine installed; condition:

Installation and feeding seed/waste materials to/from the machine description:

Any problems encountered with operation of the machine:

Procedure: Using a seed lot, set up and operate the conveyor as described. Provide the following information or answers to questions.

1. Seed crop and variety.

   Original seed lot weight.

2. Make and model of conveyor.

   Length

   Belt speed

   Seed conveying distance

   Describe conveying ribs, etc., on belt

3. Is a maintenance manual available, or an operator's manual with a detailed description of recommended maintenance? Is the machine in good condition and well maintained? What changes/improvements in maintenance would improve it?

4. How are seed fed into the conveyor?

5. What is the discharge height from the floor or conveyor base?

6. How are discharged seed handled?

7. Do seed fall out of the conveyor during operation? Describe.

8. Capacity per hour (show your calculations).

9. Do seed get underneath the conveying belt? Is a means provided for them to get out or be removed? Describe.

10. Is dust generated by the conveying action? By feeding? Discharging? Describe.

11. What are the uses and limitations of this conveyor in seed conditioning?

12. Points, conditions, or elements of special interest.

chapter twenty-four

Drag chain conveyor characteristics

Conditioner name:

Date:

Seed conditioning plant/machine location:

Machine brand, model number, other identification:

Year machine installed; condition:

Installation and feeding seed/waste materials to/from the machine description:

Any problems encountered with operation of the machine:

Procedure: Study the conveyor and provide the following information or answers to questions.

1. Make and model.

2. Name and address of manufacturer.

3. Is an operator's manual available? What is the machine designed to do? Is it being used for its designed purpose? If not, how is it being used? Can its use be changed to take better advantage of its design features? Is another machine(s) needed to improve cleaning performance?

4. Overall dimensions:

   Length

   Width

   Height

   Height, floor to seed intake and to seed discharge

   Distance seed is conveyed

   Conveying flight width and height

   Conveying trough depth

5. a. Type of conveying chain(s) used.

   b. Can links of the chain be readily replaced? Describe.

6. What are the conveyor's requirements for

   a. Power (including RPM)?

   b. Mounting and installation?

   c. Feeding seed to the conveyor?

   d. Handling seed discharging from the conveyor?

7. Describe the principle and mechanism by which seed is conveyed.

8. What are the usual uses of this conveyor in seed conditioning? Are other uses possible?

9. Does the conveying chain run in only one direction, or can it run in either direction? Describe.

10. Is the discharge height adjustable? How?

    Maximum

    Minimum

11. What is the maximum inclination at which it can effectively convey seed?

12. Describe the feed hopper that receives seed into the conveyor. Does it spill many seed?

13. Describe the discharge spouting of the conveyor.

14. Is the conveyor portable? If so, how can it be moved? Describe its undercarriage.

15. Do seed fall out of the conveyor? (Do not consider seed lost due to inadequate feed or discharge handling). How? How much? Describe.

16. Would the conveyor be difficult to clean out? Why?

17. a. Describe the drive system of the conveyor.

    b. Sketch the drive system. Show diameter and RPM of all pulleys.

18. How is the conveying chain tightened?

19. Briefly discuss and describe other makes, models, and kinds of drag chain conveyors available.

20. Points, conditions, or elements of special interest.

21. In the notebook that carries your study manual and your own observations, attach a copy of the operation and maintenance manuals of this model of machine that is installed in your facility.

22. Identify and describe five factors or conditions that influence the performance of this machine.

23. Where is seed handled by this conveyor coming from, and where is it being fed to?

24. What additional facilities, structures, subsidiary equipment, etc., are required for this machine to operate effectively and efficiently?

chapter twenty-five

Drag chain conveyor operation

Conditioner name:

Date:

Seed conditioning plant/machine location:

Machine brand, model number, other identification:

Year machine installed; condition:

Installation and feeding seed/waste materials to/from the machine description:

Any problems encountered with operation of the machine:

Procedure: Using a seed lot, set up and operate the drag chain conveyor as described. Provide the following information or answers to questions.

1. Seed crop and variety.

   Original seed lot:

      Weight

      Condition

2. Make and model of conveyor.

   Distance seed is conveyed.

   Height seed is conveyed.

3. Is a maintenance manual available, or an operator's manual with a detailed description of recommended maintenance? Is the machine in good condition and well maintained? What changes/improvements in maintenance would improve it?

4. How are seed fed into the conveyor?

5. How are seed handled as they discharge from the conveyor?

6. Capacity per hour (show calculations).

7. Are seed damaged by the conveyor? Describe.

8. Are seed lost from the conveyor? Describe.

9. Is dust created? Describe.

10. Is the conveyor difficult to clean out? Describe.

11. What are the uses and limitations of this conveyor in seed conditioning?

12. Points, conditions, or elements of special interest.

chapter twenty-six

Airlift elevator characteristics

Conditioner name:

Date:

Seed conditioning plant/machine location:

Machine brand, model number, other identification:

Year machine installed; condition:

Installation and feeding seed/waste materials to/from the machine description:

Any problems encountered with operation of the machine:

Procedure: Using a seed lot, set up and operate the airlift elevator as described. Provide the following information or answers to questions.

1. Describe the principle and mechanism by which seed is conveyed.

2. Is an operator's manual available? What is the machine designed to do? Is it being used for its designed purpose? If not, how is it being used? Can its use be changed to take better advantage of its design features? Is another machine(s) needed to improve cleaning performance?

3. How is seed conveyed in the system? What is the stated feet per minute flow speed of the conveying air stream?

4. Are there limitations on the size of seed/particles conveyed?

5. Can the conveying action be choked up easily by overfeeding? Or, does the machine's suction action created by air flow automatically prevent choking up or jamming?

6. What are installation requirements for the system? Are there any limitations on air intake?

7. Where is seed handled by this elevator coming from, and where is it being fed to?

8. What additional facilities, structures, subsidiary equipment, etc., are required for this machine to operate effectively and efficiently?

9. Is dust generated by the conveying air flow? Is a dust collector installed on the exhaust or outlet for the conveying air?

10. How is seed removed from the conveying air flow?

11. Is the settling chamber where seed is removed from the air flow lined with rubber? Are seed damaged by conveying or settling out of the air flow? Describe.

12. Is there an adequate air flow seal and seed discharge mechanism installed below the settling chamber outlet where conveyed seed is discharged? Describe its operation.

13. Can seed be taken in, or discharged out, at more than one point? Describe.

14. Is the noise level of the system a problem in the neighborhood of the plant?

15. Does the air flow system cause mechanical damage to seed? Are there any limitations on kind of seed that can be handled?

16. Overall dimensions:

    a. Length of conveying tube

    b. Diameter of conveying tube

    c. Number and angle of bends in the conveying tube

    d. Distance seed is conveyed

17. How is air flow in the conveying tube created? Can it be varied? Describe air flow generation system.

18. Can seed be discharged at more than one point? Or, can a multi-outlet distributor head be installed beneath the elevator's discharge outlet?

19. Is the inside of the seed settling chamber readily accessible for inspection and clean-out?

20. What are the elevator's requirements for

    a. Power (including RPM)?

    b. Mounting and installation?

    c. Feeding seed to the conveyor?

    d. Handling seed discharging from the elevator?

21. What are the usual uses of this elevator in seed conditioning? Are other uses possible?

22. Describe the feed hopper that receives seed into the elevator. Does it spill many seed?

23. Describe the discharge spouting of the elevator.

24. Would the elevator be difficult to clean out? Why?

25. Briefly discuss and describe other makes, models, and kinds of airlift elevators available.

26. Points, conditions, or elements of special interest.

27. In the notebook that carries your study manual and your own observations, attach a copy of the operation and maintenance manuals of this model of machine that is installed in your facility.

28. Identify and describe five factors or conditions that influence the performance of this machine.

chapter twenty-seven

Airlift elevator operation

Conditioner name:

Date:

Seed conditioning plant/machine location:

Machine brand, model number, other identification:

Year machine installed; condition:

Installation and feeding seed/waste materials to/from the machine description:

Any problems encountered with operation of the machine:

Procedure: Study the airlift elevator, noting the following:

1. Is a maintenance manual available, or an operator's manual with a detailed description of recommended maintenance? Is the machine in good condition and well maintained? What changes/improvements in maintenance would improve it?

2. Convey a lot of seed. Examine, describe, and evaluate

   a. Feeding system

   b. Conveying system

   c. Discharge system

   d. Clean-out of the system when changing varieties.

3. For this installation, what are the advantages and disadvantages of the system.

4. Points, conditions, or elements of special interest.

chapter twenty-eight

Ear corn conveyor characteristics

Conditioner name:

Date:

Seed conditioning plant/machine location:

Machine brand, model number, other identification:

Year machine installed; condition:

Installation and feeding seed/waste materials to/from the machine description:

Any problems encountered with operation of the machine:

Procedure: Using a lot of seed still on the ears (unshelled), use the ear maize conveyor and observe its operating characteristics.

1. Is an operator's manual available? What is the machine designed to do? Is it being used for its designed purpose? If not, how is it being used? Can its use be changed to take better advantage of its design features? Is another machine(s) needed to improve cleaning performance?

2. Does the conveyor move the ears gently, without shelling and losing much seed, or damaging seed?

3. Points, conditions, or elements of special interest.

4. In the notebook that carries your study manual and your own observations, attach a copy of the operation and maintenance manuals of this model of machine that is installed in your facility.

5. Identify and describe five factors or conditions that influence the performance of this machine.

chapter twenty-nine

Ear corn conveyor operation

Conditioner name:

Date:

Seed conditioning plant/machine location:

Machine brand, model number, other identification:

Year machine installed; condition:

Procedure: Study the ear corn conveyor, noting the following:

1. Is a maintenance manual available, or an operator's manual with a detailed description of recommended maintenance? Is the machine in good condition and well maintained? What changes/improvements in maintenance would improve it?

2. Installation and feeding seed/waste materials to/from the machine description.

3. Any problems encountered with operation of the machine?

4. Points, conditions, or elements of special interest.

chapter thirty

Corn (maize) sheller characteristics

Conditioner name:

Date:

Seed conditioning plant/machine location:

Machine brand, model number, other identification:

Year machine installed; condition:

Installation and feeding seed/waste materials to/from the machine description:

Any problems encountered with operation of the machine:

Procedure: Study the sheller and answer the following:

1. What is the purpose of the corn (maize) sheller?

2. On what crop seed is the corn (maize) sheller used?

3. From where and in what condition does seed come to the sheller? Where does the seed go when it leaves the sheller?

4. What is the optimum corn (maize) seed moisture content for shelling? Why?

5. At what point in the corn seed conditioning line or sequence is the sheller normally used?

6. Is an operator's manual available? What is the machine designed to do? Is it being used for its designed purpose? If not, how is it being used? Can its use be changed to take better advantage of its design features? Is another machine(s) needed to improve cleaning performance?

7. Describe the usual installation and operation of a sheller in the corn seed conditioning plant.

8. Make and model.

9. Name and address of manufacturer.

10. What mechanism and action does this sheller use to remove corn seed from cobs? Describe.

11. Draw a simple diagram to illustrate the flow of ear corn into the sheller; the shelling operation; and discharge of shelled corn seed, cobs, and dusty air. Label all essential parts.

12. What are corn sheller's requirements for

    a. Power (including RPM)?

    b. Installation and mounting?

    c. Handling cobs?

d. Handling shelled corn seed?

e. Dust control?

13. Overall dimensions:

Height

Width

Length

Height, floor to ear corn intake

Height, floor to shelled corn seed discharge

Height, floor to cob discharge

14. Draw the mounting base of the corn sheller (i.e., the part that sits on the floor or frame). Show all dimensions and the location of the ear corn intake, cob discharge, shelled corn seed discharge, and dust exhaust.

15. List and describe all adjustments on the corn sheller.

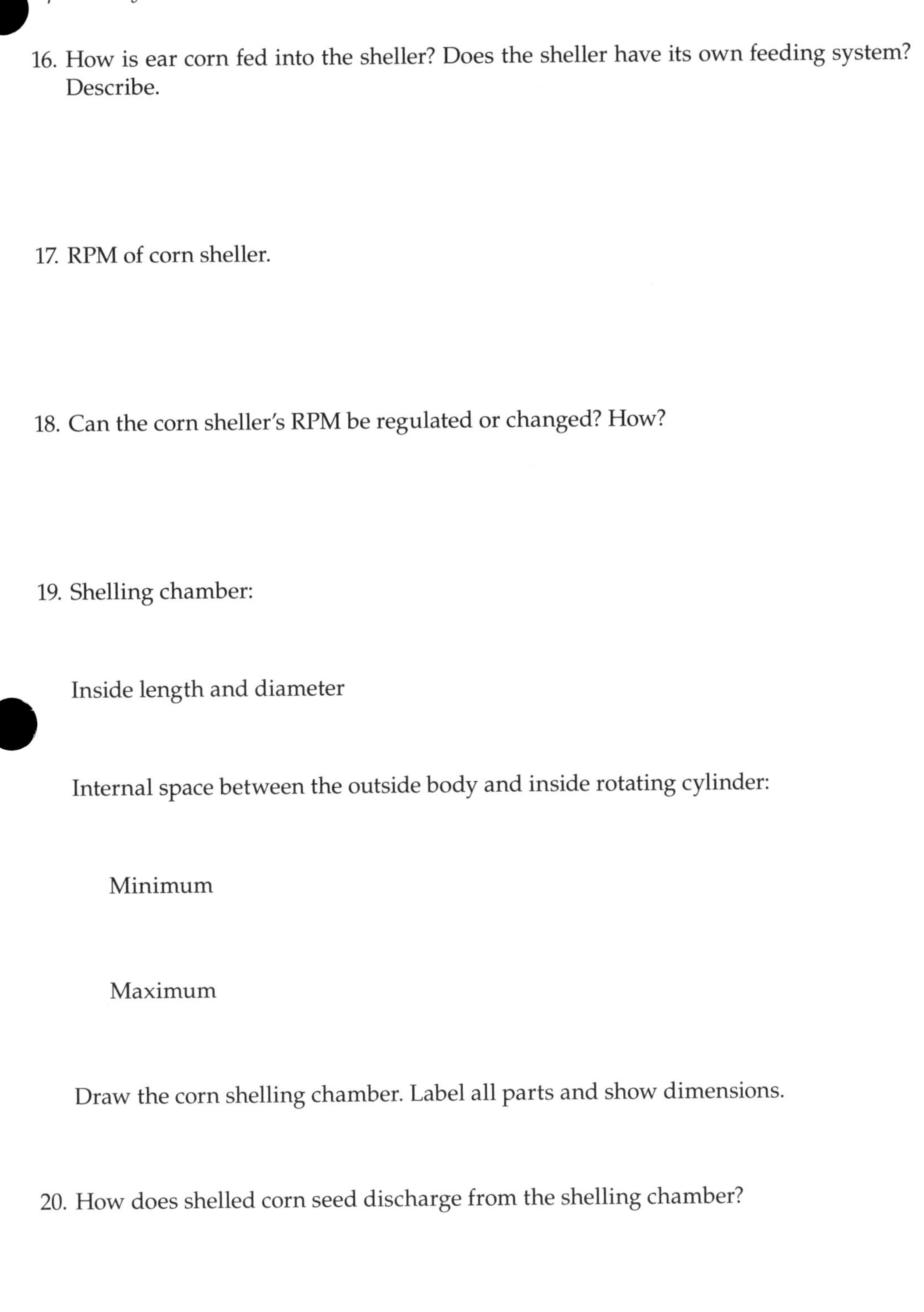

16. How is ear corn fed into the sheller? Does the sheller have its own feeding system? Describe.

17. RPM of corn sheller.

18. Can the corn sheller's RPM be regulated or changed? How?

19. Shelling chamber:

    Inside length and diameter

    Internal space between the outside body and inside rotating cylinder:

    Minimum

    Maximum

    Draw the corn shelling chamber. Label all parts and show dimensions.

20. How does shelled corn seed discharge from the shelling chamber?

21. Is the shelled corn seed cleaned by air blast, screens, etc., in the sheller? Where? Describe the mechanism and its action.

22. Is shelled corn seed separated from the discharging cobs leaving the shelling chamber? How? Where? Describe in detail.

23. Do cobs fall directly out of the sheller, or is there a built-in mechanism to blow or convey cobs to some distance away from the sheller? Describe.

24. a. How, and at what points, is dust removed from the sheller? Describe.

    b. Draw the dust removal system of the corn sheller.

25. Describe the sequence of initial adjustments for setting up and operating the corn sheller.

26. Briefly discuss and describe other makes, models, and kinds of corn shellers available.

27. Points, conditions, or elements of special interest.

28. In the notebook that carries your study manual and your own observations, attach a copy of the operation and maintenance manuals of this model of machine that is installed in your facility.

29. Identify and describe five factors or conditions that influence the performance of this machine.

chapter thirty-one

Corn sheller operation

Conditioner name:

Date:

Seed conditioning plant/machine location:

Machine brand, model number, other identification:

Year machine installed; condition:

1. Is a maintenance manual available, or an operator's manual with a detailed description of recommended maintenance? Is the machine in good condition and well maintained? What changes/improvements in maintenance would improve it?

Installation and feeding seed/waste materials to/from the machine description:

Any problems encountered with operation of the machine:

Procedure: Using a seed lot, set up, adjust, and operate the corn sheller as described. Provide the following information or answers to questions.

1. Variety:

   Original:

   Weight

   % Moisture content

   % Germination

   Test weight when hand shelled (e.g., pounds per bushel weight)

   Mechanical injury % when hand shelled

2. Make and model of sheller.

3. Approximate adjustments used.

Adjustment	Setting used

4. Describe the condition of cobs leaving the sheller.

5. Are many unshelled seed left on cobs? Describe. Remove seed by hand from a weighed sample of shelled-out cobs and determine amount left on the cobs.

6. Examine a sample of shelled corn seed under magnification. Describe the % of damaged seed and type of mechanical damage to seed.

7. Approximate capacity per hour.

8. Shelled corn seed weight and % of the original unshelled seed lot weight. Determine weight of cobs and % of original seed lot weight. Determine how much is the "invisible" weight loss due to dust, weight of seed material left in sheller, etc.

9. Make significant changes in adjustments, as listed below, and note the results on operation, seed condition, mechanical damage, and weight of each discharged fraction. (After noting an adjustment, return it to normal operation before changing the next adjustment.)

   a. Decrease feed rate

   b. Increase feed rate

   c. Decrease sheller speed (if possible).

d. Increase sheller speed (if possible)

e. Increase air

f. Decrease air

10. Describe the dust created, and handling of cobs. How would you suggest that a corn sheller be installed?

11. Points, conditions, or elements of special interest.

chapter thirty-two

Scalper (precleaner) characteristics

Conditioner name:

Date:

Seed conditioning plant/machine location:

Machine brand, model number, other identification:

Year machine installed; condition:

Installation and feeding seed/waste materials to/from the machine description:

Any problems encountered with operation of the machine:

Procedure: Study the scalper and provide the following information or answers to questions.

1. What is the purpose of the scalper?

2. On what crop seed is the scalper used? Why?

3. Where does the seed go after it leaves the scalper?

4. Principle(s) by which the scalper separates seed.

5. Is an operator's manual available? What is the machine designed to do? Is it being used for its designed purpose? If not, how is it being used? Can its use be changed to take better advantage of its design features? Is another machine(s) needed to improve cleaning performance?

6. Components, structures, etc., required for complete and efficient operation.
   A study of establishing and equipping a seed conditioning plant shows that scalpers usually require the following. In most cases, the manufacturer/supplier will provide recommendations, including plans.

Component, etc.	Needed for this machine	Comments
a. Elevator bringing seed not previously conditioned on this machine to the holding bin serving this machine. This elevator should have a two-way valve on its discharge, so that seed may bypass this machine if it is not needed.	Needed	To fit the machine's installation
b. Spouting of seed (coming from the previous separator) to this elevator.	Needed	To fit the machine's installation
c. Holding bin mounted over the feed intake of this machine.	Needed	To fit the machine's installation
d. Spouting of seed from this elevator to the holding bin mounted over this machine.	Needed	To fit the machine's installation
e. Spouting or flow of seed from the bin into the feed intake of this machine.	Needed	To fit the machine's installation
f. Mounting frame to support this machine in a position and at a height required for efficient operation.	Needed	To fit the machine's installation
g. Worker platforms, with required access steps or ladders, in appropriate locations so that operators can easily reach all controls of the machine for adjustments, and all parts of the machine for clean-out, setup, and repair.	Needed	To fit the machine's installation
h. Spouting or flow system to take good seed leaving this machine to the bin or conveyor/elevator taking good seed to the next machine in the conditioning sequence.	Needed	To fit the machine's installation
i. Dust and dusty air exhaust ducting and outside collector(s) handling dusty air and/or light materials separated from the good seed by this machine.	As needed	If needed, to fit the machine's installation
j. If the air around this machine is dusty, an exhaust fan/ducting/collection system should be installed to remove all dusty air.	As needed	If needed, to fit the machine's installation
k. Spouting to take each fraction of removed undesirable materials either to (1) bag collecting this material at this conditioning machine or to (2) a conveyor system taking all fractions of separated undesirable materials to a central waste product collecting system.	Needed	To fit the machine's installation
l. Elevator and required spouting to collect middling fraction(s) discharged from this machine back into the good-seed bin serving this machine, for reconditioning.	Not usually needed	Not needed

7. Describe the separation(s) made by the scalper; that is, how and what it does.

8. Where would a scalper be installed and used in the conditioning line or sequence?

9. What conditioning, if any, is usually done to a seed lot before it goes to the scalper?

10. Make and model of the scalper.

11. Name and address of manufacturer.

12. Illustrate, with a simple diagram, the flow of seed into and through the scalper, and all discharged fractions. Label all discharged fractions and all essential parts of the scalper.

13. Overall dimensions:

    Height

    Width

Length

Height, floor to seed feed intake

Height, floor to clean seed discharge

14. What are the scalper's requirements for?

    a. Power (including RPM)?

    b. Installation and mounting?

    c. Seed feed and supply?

    d. Handling discharged clean seed?

    e. Handling discharged waste materials?

f. Air ducting and dust control?

g. What kind of screen(s) does the scalper use?

h. How many screens of each type

i. Screen length and width

15. Does the scalper also use an air separation? Describe.

16. Draw the scalper's mounting base (i.e., part which sits on the floor or mounting frame). Show all dimensions, as well as the position of seed feed intake and all discharge spouts.

17. List and describe adjustments that can be made to affect the separation.

18. Describe the scalper's feeding mechanism. Onto how much of the screen's width are seed fed? Is there a mechanism to ensure uniform feeding of trashy seed? Describe.

19. Can seed feed be stopped while the scalper is still running? How?

20. How can the feed hopper be cleaned out?

21. Can screens be readily changed? Describe.

22. Are screen brushes or other screen cleaning system used? Describe the mechanism used to prevent jamming and plugging of screen perforations.

23. Can screen pitch be adjusted? How? How much?

24. Can screen oscillation (vibration) be adjusted? How? How much?

25. How is overall undesirable vibration of the machine minimized?

26. Describe the sequence of initial adjustments for setting up and operating the scalper.

27. Briefly discuss and describe other makes, models, and kinds of scalper available.

28. Points, conditions, or elements of special interest.

29. In the notebook that carries your study manual and your own observations, attach a copy of the operation and maintenance manuals of this model of machine that is installed in your facility.

30. Identify and describe five factors or conditions which influence the performance of this machine.

chapter thirty-three

Scalper operation

Conditioner name:

Date:

Seed conditioning plant/machine location:

Machine brand, model number, other identification:

Year machine installed; condition:

Installation and feeding seed/waste materials to/from the machine description:

Any problems encountered with operation of the machine:

Procedure: Using a seed lot, set up, adjust, and operate the scalper as described.

1. Seed crop and variety, original % purity, % germination, test weight, purpose of using the scalper, conditioning previously done on the seed lot

2. Approximate adjustments used.

Adjustment	Setting used

3. Is a maintenance manual available, or an operator's manual with a detailed description of recommended maintenance? Is the machine in good condition and well maintained? What changes/improvements in maintenance would improve it?

4. Separation results.

Spout	Material discharged	Test weight	% Pure seed	% Germination	Weight	% of original

5. Was the desired separation accomplished? How has the seed lot been improved?

6. Were many good seed lost with the separated waste materials? How could this loss be reduced? How could the lost good seed be recovered?

7. Make significant changes in adjustments as indicated below, and note the results. (After noting an adjustment, return it to normal operation before changing the next adjustment.)

    a. Increase feed rate

    b. Decrease feed rate

    c. Increase air

    d. Decrease air

    e. Increase screen vibration speed

    f. Decrease screen vibration speed

g. Increase feed rate

h. Decrease feed rate

i. Change positions of screens

8. Points, conditions, or elements of special interest.

chapter thirty-four

Debearder characteristics

Conditioner name:

Date:

Seed conditioning plant/machine location:

Machine brand, model number, other identification:

Year machine installed; condition:

Installation and feeding seed/waste materials to/from the machine description:

Any problems encountered with operation of the machine:

Procedure: Study the debearder and provide the following information or answers to questions.

1. For what purposes is the debearder used in seed conditioning?

2. On what crop seed is the debearder used? Why?

3. Where does the crop seed go after it leaves the debearder?

4. Describe any partial conditioning usually done to seed before going to the debearder.

5. Is an operator's manual available? What is the machine designed to do? Is it being used for its designed purpose? If not, how is it being used? Can its use be changed to take better advantage of its design features? Is another machine(s) needed to improve cleaning performance?

6. How and where in the seed conditioning line or sequence is the debearder usually installed? Describe.

7. Make and model of debearder.

8. Name and address of manufacturer.

9. How is the debearding operation performed?

10. Diagram the flow of seed into and through the debearder as it is debearded and then discharges from the machine. Label all essential parts of the debearder.

11. Overall dimensions:

    Length

    Width

    Height

    Height, floor to seed feed intake

    Height, floor to seed discharge

Component, etc.	Needed for this machine	Comments
a. Elevator bringing seed not previously conditioned on this machine to the holding bin serving this machine. This elevator should have a two-way valve on its discharge, so that seed may bypass this machine if it is not needed.	Needed	To fit the machine's installation
b. Spouting of seed (coming from the previous separator) to this elevator.	Needed	To fit the machine's installation
c. Holding bin mounted over the feed intake of this machine.	Needed	To fit the machine's installation
d. Spouting of seed from this elevator to the holding bin mounted over this machine.	Needed	To fit the machine's installation
e. Spouting or flow of seed from the bin into the feed intake of this machine.	Needed	To fit the machine's installation
f. Mounting frame to support this machine in a position and at a height required for efficient operation.	Needed	To fit the machine's installation
g. Worker platforms, with required access steps or ladders, in appropriate locations so that operators can easily reach all controls of the machine for adjustments, and all parts of the machine for clean-out, setup, and repair.	Needed	To fit the machine's installation
h. Spouting or flow system to take good seed leaving this machine to the bin or conveyor/elevator taking good seed to the next machine in the conditioning sequence.	Needed	To fit the machine's installation
i. Dust and dusty air exhaust ducting and outside collector(s) handling dusty air and/or light materials separated from the good seed by this machine.	As needed	If needed, to fit the machine's installation
j. If the air around this machine is dusty, an exhaust fan/ducting/collection system should be installed to remove all dusty air.	As needed	If needed, to fit the machine's installation
k. Spouting to take each fraction of removed undesirable materials either to (1) bag collecting this material at this conditioning machine or to (2) a conveyor system taking all fractions of separated undesirable materials to a central waste product collecting system.	Needed	To fit the machine's installation
l. Elevator and required spouting to collect middling fraction(s) discharged from this machine back into the good-seed bin serving this machine, for reconditioning.	Not usually needed	Not needed

12. What are the debearder's requirements for

    a. Power (including RPM)?

    b. Installation and mounting?

    c. Dust control?

    d. Seed supply and feed?

    e. Handling discharged seed?

13. Draw the debearder's mounting base (i.e., the part which sits on the floor or installation frame). Label all dimensions and show location of seed intake, seed discharge, and air duct outlet.

14. List and describe all adjustments on the debearder.

15. How are seed fed into the debearder? Can seed feed be stopped while the debearder is still operating? How?

16. How does changing the rate of seed feed affect the debearding action?

17. Inside dimensions of the body of the debearder:

    Diameter

    Length

18. Length of the revolving shaft that carries the beater arms.

    Beater arms on the revolving shaft:

    Number

    Length

    Angle of installation

19. Stationary beater arms mounted on debearder body:

    Number of rows

    Number of beater arms

    Length of beater arms

    Angle of beater arms

20. Distance or space between stationary and rotating beater arms as the shaft turns.

21. Can RPM of beater shaft be varied? How?

22. What is beater shaft's minimum RPM and maximum RPM?

23. Dimension of the seed discharge gate:

    Width

    Height

24. Describe the use and purpose of the movable weight on the hinged seed discharge gate.

25. Does the debearder have a dust exhaust system? Describe.

26. The dust control system exhausts air and dust from what points?

27. How is the debearder cleaned out?

28. Describe the sequence of initial adjustments for setting up and operating the debearder.

29. Briefly discuss and describe other makes, models, and kinds of debearders available.

30. Points, conditions, or elements of special interest.

31. In the notebook that carries your study manual and your own observations, attach a copy of the operation and maintenance manuals of this model of machine that is installed in your facility.

32. Identify and describe five factors or conditions that influence the performance of this machine.

chapter thirty-five

Debearder operation

Conditioner name:

Date:

Seed conditioning plant/machine location:

Machine brand, model number, other identification:

Year machine installed; condition:

Installation and feeding seed/waste materials to/from the machine description:

Any problems encountered with operation of the machine:

Procedure: Using a suitable seed lot, set up, adjust, and operate the debearder as described. Provide the following information or answers to questions.

1. Seed crop and variety.

   Seed lot weight, % of germination, % purity

   Purpose of debearding

   Previous conditioning of this seed lot

2. Make, model, and manufacturer of the debearder.

3. Is a maintenance manual available, or an operator's manual with a detailed description of recommended maintenance? Is the machine in good condition and well maintained? What changes/improvements in maintenance would improve it?

4. Approximate adjustments used.

Adjustment	Setting used

5. Debearding results:

   a. Appearance of seed, including removal of awns, etc.

   b. Has flowability and plantability of seed been improved? Describe.

   c. Test weight and % weight loss.

   d. Percentage of mechanically damaged seed; describe types of damage.

   e. Approximate capacity per hour.

6. Make significant changes in adjustments as indicated below, and note the results. (After noting an adjustment, return it to normal operation before making the next change.)

   a. Increase feed rate

   b. Decrease feed rate

c. Increase speed of beater shaft

d. Decrease speed of beater shaft

e. Move weight on discharge gate toward the gate (less resistance to seed discharge)

f. Move weight on discharge gate away from gate (more resistance to seed discharge)

g. Increase dust exhaust air

h. Decrease dust exhaust air

chapter thirty-six

Huller-scarifier characteristics

Conditioner name:

Date:

Seed conditioning plant/machine location:

Machine brand, model number, other identification:

Year machine installed; condition:

Installation and feeding seed/waste materials to/from the machine description:

Any problems encountered with operation of the machine:

Procedure: Study the huller-scarifier and provide the following information or answers to questions.

1. What is the purpose of the huller-scarifier? What does it do to seed?
2. On what crop seed is the huller-scarifier used? Why?
3. Where does the crop seed come from as it goes to the huller-scarifier?
4. Where does the crop seed go as it leaves the huller-scarifier?
5. How and where is the huller-scarifier usually installed in the conditioning line or flow sequence? Describe.
6. Is an operator's manual available? What is the machine designed to do? Is it being used for its designed purpose? If not, how is it being used? Can its use be changed to take better advantage of its design features? Is another machine(s) needed to improve cleaning performance?
7. Describe any partial conditioning normally done to seed before using the huller-scarifier. Why is it done?

8. Make and model.

9. Name and address of manufacturer.

Component, etc.	Needed for this machine	Comments
a. Elevator bringing seed not previously conditioned on this machine to the holding bin serving this machine. This elevator should have a two-way valve on its discharge, so that seed may bypass this machine if it is not needed.	Needed	To fit the machine's installation
b. Spouting of seed (coming from the previous separator) to this elevator.	Needed	To fit the machine's installation
c. Holding bin mounted over the feed intake of this machine.	Needed	To fit the machine's installation
d. Spouting of seed from this elevator to the holding bin mounted over this machine.	Needed	To fit the machine's installation
e. Spouting or flow of seed from the bin into the feed intake of this machine.	Needed	To fit the machine's installation
f. Mounting frame to support this machine in a position and at a height required for efficient operation.	Needed	To fit the machine's installation
g. Worker platforms, with required access steps or ladders, in appropriate locations so that operators can easily reach all controls of the machine for adjustments, and all parts of the machine for clean-out, setup, and repair.	Needed	To fit the machine's installation
h. Spouting or flow system to take good seed leaving this machine to the bin or conveyor/elevator taking good seed to the next machine in the conditioning sequence.	Needed	To fit the machine's installation
i. Dust and dusty air exhaust ducting and outside collector(s) handling dusty air and/or light materials separated from the good seed by this machine.	As needed	If needed, to fit the machine's installation
j. If the air around this machine is dusty, an exhaust fan/ducting/collection system should be installed to remove all dusty air.	As needed	If needed, to fit the machine's installation
k. Spouting to take each fraction of removed undesirable materials either to (1) bag collecting this material at this conditioning machine or to (2) a conveyor system taking all fractions of separated undesirable materials to a central waste product collecting system.	Needed	To fit the machine's installation
l. Elevator and required spouting to collect middling fraction(s) discharged from this machine back into the good-seed bin serving this machine, for reconditioning.	Not usually needed	Not needed

10. How are seed hulled and/or scarified in this machine? Describe the operation and parts or components that accomplish it.

11. Draw a simple diagram illustrating the flow of seed into and through the huller-scarifier. Label all essential components.

    a. Where is seed handled by this machine coming from, and where is it being fed to?

    b. What additional facilities, structures, subsidiary equipment, etc., are required for this machine to operate effectively and efficiently?

12. Overall dimensions:

    Height

    Width

    Length

    Height, floor to seed feed intake

    Height, floor to seed discharge

13. What are the huller-scarifier's requirements for

    a. Power (including RPM)?

    b. Installation and mounting?

    c. Seed supply and feed?

    d. Air ducting and dust control?

    e. Handling discharged seed?

14. Draw the mounting base (i.e., the part that sits on the floor or mounting frame) of the huller-scarifier. Show all dimensions, as well as the position of seed feed intake, seed discharge spout, and air duct outlets.

15. List and describe all adjustments on the huller-scarifier.

16. Describe how seed feed to the huller-scarifier is adjusted, and how the adjustment works. How does feed rate affect the machine's actions?

17. a. Minimum and maximum RPM of the huller-scarifier.

    b. Describe the shape, dimensions, material, and construction of the hulling-scarifying discs or cylinders.

18. Can the hulling/scarifying mechanism be changed, to modify its action? Describe.

19. How are hulling/scarifying discs (or cylinders) changed? How much time is required?

20. Dust exhaust system:

    Type of fan

    Number of fan blades

    Fan diameter

    Fan RPM

21. How is air flow adjusted?

22. Describe the sequence of initial adjustments for setting up and operating the huller-scarifier.

23. Briefly discuss and describe other makes, models, and types of huller-scarifiers available.

24. Points, conditions, or elements of special interest.

25. In the notebook that carries your study manual and your own observations, attach a copy of the operation and maintenance manuals of this model of machine that is installed in your facility.

26. Identify and describe five factors or conditions that influence the performance of this machine.

chapter thirty-seven

Huller-scarifier operation

Conditioner name:

Date:

Seed conditioning plant/machine location:

Machine brand, model number, other identification:

Year machine installed; condition:

Installation and feeding seed/waste materials to/from the machine description:

Any problems encountered with operation of the machine:

Procedure: Using a suitable seed lot, set up, adjust, and operate the huller-scarifier as described. Provide the following information or answers to questions.

1. Seed crop and variety.

   Original seed lot:

   Weight

   Test weight

   % Germination

   % Hard seed

   % Purity

2. Purpose of using the huller-scarifier.

   a. Is a maintenance manual available, or an operator's manual with a detailed description of recommended maintenance? Is the machine in good condition and well maintained? What changes/improvements in maintenance would improve it?

3. Previous conditioning of this seed lot.

4. Make, model, and manufacturer of the huller-scarifier.

5. Approximate adjustments used.

Adjustment	Setting used

6. Examine, under magnification, samples of seed both before and after being hulled/scarified. Describe the results of hulling/scarifying.

7. Approximate capacity per hour.

8. After hulling/scarifying:

   Final seed lot weight

   % Weight lost

   Test weight

   % Purity

9. Has the seed lot been improved? How?

10. What further conditioning is necessary?

11. Make significant changes in adjustments as indicated below, and note the results. (After noting an adjustment, return it to normal operation before changing the next adjustment.)

    a. Increase feed rate

    b. Decrease feed rate

    c. Increase speed of hulling/scarifying disc or cylinder

    d. Decrease speed of hulling/scarifying disc or cylinder

    e. Decrease air

f. Increase air

g. Change type of hulling-scarifying disc/cylinder

12. Points, conditions, or elements of special interest.

chapter thirty-eight

Air-screen cleaner characteristics

Conditioner name:

Date:

Seed conditioning plant/machine location:

Machine brand, model number, other identification:

Year machine installed; condition:

Installation and feeding seed/waste materials to/from the machine description:

Any problems encountered with operation of the machine:

Procedure: Study the air-screen cleaner and provide the following information or answers to questions.

1. Seed separating principle(s) used.

2. What are the usual uses of the air-screen cleaner in seed conditioning? What other uses are possible? Why is the air-screen cleaner the basic step in seed conditioning?

3. What is the usual position of the air-screen cleaner in the conditioning line or sequence?

4. From where does the crop seed come as it goes to the air-screen cleaner?

5. On what crop seed is the air-screen cleaner used? Why?

6. Where does each crop seed go as it leaves the air-screen cleaner?

7. Is an operator's manual available? What is the machine designed to do? Is it being used for its designed purpose? If not, how is it being used? Can its use be changed to take better advantage of its design features? Is another machine(s) needed to improve cleaning performance?

8. Where is seed handled by this machine coming from, and where is it being fed to?

Component, etc.	Needed for this machine	Comments
a. Elevator bringing seed not previously conditioned on this machine to the holding bin serving this machine. This elevator should have a two-way valve on its discharge, so that seed may bypass this machine if it is not needed.	Needed	To fit the machine's installation
b. Spouting of seed (coming from the previous separator) to this elevator.	Needed	To fit the machine's installation
c. Holding bin mounted over the feed intake of this machine.	Needed	To fit the machine's installation
d. Spouting of seed from this elevator to the holding bin mounted over this machine.	Needed	To fit the machine's installation
e. Spouting or flow of seed from the bin into the feed intake of this machine.	Needed	To fit the machine's installation
f. Mounting frame to support this machine in a position and at a height required for efficient operation.	Needed	To fit the machine's installation
g. Worker platforms, with required access steps or ladders, in appropriate locations so that operators can easily reach all controls of the machine for adjustments, and all parts of the machine for clean-out, setup, and repair.	Needed	To fit the machine's installation
h. Spouting or flow system to take good seed leaving this machine to the bin or conveyor/elevator taking good seed to the next machine in the conditioning sequence.	Needed	To fit the machine's installation
i. Dust and dusty air exhaust ducting and outside collector(s) handling dusty air and/or light materials separated from the good seed by this machine.	As needed	If needed, to fit the machine's installation
j. If the air around this machine is dusty, an exhaust fan/ducting/collection system should be installed to remove all dusty air.	As needed	If needed, to fit the machine's installation
k. Spouting to take each fraction of removed undesirable materials either to (1) bag collecting this material at this conditioning machine or to (2) a conveyor system taking all fractions of separated undesirable materials to a central waste product collecting system.	Needed	To fit the machine's installation
l. Elevator and required spouting to collect middling fraction(s) discharged from this machine back into the good-seed bin serving this machine, for reconditioning.	Not usually needed	Not needed

9. What additional facilities, structures, subsidiary equipment, etc., are required for this machine to operate effectively and efficiently?

10. What conditioning is usually done to a seed lot before it goes to the air-screen cleaner? Why?

11. What additional conditioning is, or may be, done to a seed lot after cleaning on the air-screen cleaner? Describe principles and procedures.

12. Describe how the air-screen cleaner separates seed.

13. Make and model of air-screen cleaner examined.

14. Name and address of manufacturer.

15. Overall dimensions:

    Height

    Width

Length

Height, floor to seed feed intake

Height, floor to good seed discharge

16. Draw the mounting base (i.e., the part which sits on the floor or on the mounting frame) of the air-screen cleaner. Show all dimensions, as well as position of the seed feed intake, good seed discharge, and all waste discharge spouts.

17. What are the air-screen cleaner's requirements for

    a. Power (including RPM)?

    b. Seed supply and feed?

    c. Installation and mounting?

    d. Operator access and work space, including changing screens?

e. Handling discharged waste products?

f. Handling discharged good seed?

g. Air ducting and dust control?

18. Draw a diagram to illustrate the flow of seed into and through the air-screen cleaner, and of all separated fractions discharged from the cleaner. Label all seed fractions and essential parts of the cleaner.

19. List and describe all adjustments, changes, etc., that can be made on the air-screen cleaner to affect its separation.

20. How should seed be supplied to the air-screen cleaner's feed hopper? Why?

21. How is the rate of feeding seed into the air-screen cleaner controlled?

22. The feed rate can be adjusted by adjusting both

    a.

    b.

    Explain.

23. How does the feed hopper feed seed into the cleaner? Where do the seed go as they leave the feed hopper? Over how much of the screen width are seed fed, and why?

24. Can the seed feed be stopped without closing the hopper or turning the cleaner off? How?

25. Why should the feed be stopped before the cleaner is turned off?

26. Describe the feed roll inside the feed hopper. What is its purpose? How does it function? Why is it constructed in this manner?

27. How is the feed roll driven? Can its speed be adjusted? How? Describe the mechanism and its action on feed rate.

28. What happens to the rate of feed when the adjustable gate above the feed roll (inside the feed hopper) is

   a. Raised?

   b. Lowered?

29. Observe the spiked steel shaft running through the feed hopper. How does it function? What purpose does it serve?

30. How is access into the feed hopper provided for clean-out/inspection?

31. What is the purpose of the adjustable gate below the feed roll, outside the feed hopper? Explain how and when it is used.

32. How can you determine that the feed rate is correct? Discuss.

33. How many screen "shoes" does this air-screen cleaner have? How many screens are in each shoe?

34. How many screens does this air-screen cleaner have? What is the length and width of the screens? What is the total screening surface area of each screen?

35. How many scalping screens does this cleaner have? How many bottom or grading screens? Can these be changed? How? Where is each located?

36. Draw a diagram showing the position of all screen shoes and screens. Label each screen as to its function (i.e., scalping or grading); show the flow of good seed.

37. a. The screen positions are numbered from ____ to ____.

    b. In cleaning a seed lot, what is the function of each screen?

        1.

        2.

        3.

        4.

5.

6.

38. Why must screens be securely fastened in the screen shoes?

39. How are screens secured in position in the screen shoes? Describe.

40. How are screens removed from, or placed in, the cleaner? Describe.

41. a. Can the pitch (angle of inclination, from horizontal) of screens be changed?

b. Which screens? How much change can be made?

c. How is the pitch of screens changed?

d. Why is the pitch of screens changed?

42. a. When would a flat screen pitch be used?

    b. When would a steep screen pitch be used?

43. Can the oscillation or vibration of the screen shoes and screens be changed? How much?

44. Describe the mechanical system that causes the screens/shoes to vibrate, and how it works.

45. a. When would a lower speed of screen vibration be used?

    b. When would a higher speed of screen vibration be used?

46. How is undesirable vibration of the total machine minimized?

47. a. What is the purpose of the brushes beneath each screen?

    b. How many brushes are under each screen?

48. Describe the movement of the brushes beneath the screen. What mechanism is used to drive the brushes? How does it function?

49. How are the brushes supported and connected to the drive cable? Remove a brush carriage and observe how it and the brushes are mounted. Describe.

50. How is the position of brushes adjusted for proper pressure against the screens?

51. Describe the proper adjustment of brushes against the under side of the screens.

52. a. What is the result of brushes being adjusted too low beneath the screens (i.e., not touching the screen)?

    b. What is the result of brushes being adjusted too high (i.e., with excessive pressure against the bottom of the screens)?

53. When brushes are installed in the cleaner, the blank space (i.e., without bristles) of the brushes must be aligned with what? How?

54. How can proper/improper alignment of brushes and screens be observed and determined?

55. Some air-screen cleaners use a system of rubber balls instead of brushes to keep screen perforations free of jammed seed.

    a. Describe the rubber ball screen-cleaning mechanism.

    b. How do the balls remove seed jammed into screen perforations?

    c. What are the advantages of the rubber balls over the brush system?

    d. What are disadvantages of the rubber balls as compared to the brush system?

56. How many fans and air separations does the cleaner have?

57. Fans:

Number

Type

RPM

Number of blades

Width of blades

Is fan RPM adjustable? How?

58. Which fan and control operate the

a. Upper air separation?

b. Lower air separation?

59. Is each air separation a distinct, independent system, or do they work together? Is it necessary to "balance" the two air separations? How? Why? Describe.

60. How and where does the upper air separation function? What does it separate?

61. How and where does the lower air separation function? What does it separate?

62. How is each air separation adjusted to the desired separating action? How can you tell when the air separation is properly adjusted?

63. When the control of an air separation is adjusted, how does it modify the air flow, thereby adjusting the separation made?

64. What is the purpose of the hinged gates on the spouts which discharge light materials taken out by the air separation? What happens if these gates are fastened in the open position?

65. Is the light material removed by the air separations blown out to the dust collector, or is it discharged from an air settling chamber and spout in the cleaner? Does the operator have a choice? Describe fully.

66. What happens when the air liftings product is not properly removed, or the air liftings discharge jams? How can you know when this begins to occur?

67. a. What is the purpose of the screen tappers that strike against the wood strips on top of the screens?

b. Why must the screen tappers strike only against the wood strips, not against the screen itself?

c. What mechanism drives the screen tappers? How does it function?

d. Can the force of screen tappers be adjusted? How?

e. How are the screen tappers disconnected or connected?

68. What is the purpose of the adjustable gate in the pan beneath screen no. 2 and the adjustable gate at the discharge end of this screen? Describe how they are adjusted for each use situation. (Note: this is not on all models.)

a.

b.

69. Where are waste products discharged from the cleaner? List and briefly describe all waste fractions and their discharge spouts and positions.

70. Can more than one waste product be combined, so that they discharge from the cleaner through a single spout? Describe.

71. Where and how are clean seed discharged from the cleaner? Describe.

72. How should dust and light trash be removed from the air exhausted from the cleaner?

73. If the air ducts and dust collectors receive undue vibration from the cleaner, how can this be eliminated?

74. Describe the initial sequence of adjustments for setting up and operating the air-screen cleaner.

75. Briefly discuss and describe other makes, models, and kinds of air-screen cleaners available.

76. Points, conditions, or elements of special interest.

77. In the notebook that carries your study manual and your own observations, attach a copy of the operation and maintenance manuals of this model of machine that is installed in your facility.

78. Identify and describe five factors or conditions which influence the performance of this machine.

chapter thirty-nine

Screens

Conditioner name:

Date:

Seed conditioning plant/machine location:

Machine brand, model number, other identification:

Year machine installed; condition:

Installation and feeding seed/waste materials to/from the machine description:

Any problems encountered with operation of the machine:

Procedure: Study the design, construction, use, and identification of screens. Provide the following information or answers to questions.

1. Why does the air-screen cleaner have removable screens, with many screens of different perforation sizes and shapes available?

2. Name and describe the three major parts of a screen.

   a. When a screen is installed in an air-screen cleaner, the screen end with a blank space (apron) always is at which end?

   b. Why? What is its purpose?

3. Why is it necessary to keep the screen surface uniformly level and even? What happens when the screen surface is not level? What should be done with screens whose surface has become uneven?

4. Draw a screen for the ______________ air-screen cleaner. Show all dimensions. Also, draw a side cutaway view to show all screen frame ribs.

5. Name and describe the two types of materials used for the perforated surfaces of screens.

6. What shapes of perforations are available in sheet metal screens?

a.

b.

c.

d.

7. What shapes of perforations are available in wire mesh screens?

a.

b.

c.

d.

8. List the different types of screen perforations, the shape of each, and explain how each is measured and identified.

a.

b.

c.

d.

e.

f.

9. a. Why is the percentage of open area (in perforations) important in a screen?

b. What is the major disadvantage of having a high % of open area in a perforated sheet metal screen? Discuss.

10. a. About how many screen perforation sizes are available?

Number of different kinds/shapes of perforations/openings.

Supplier

b. List, by type, the available screen sizes.

11. Screen perforations are constantly swept from beneath by screen brushes (some models use a different mechanism). What happens if the brushes are adjusted so that they are

a. Too high (i.e., press excessively against the screen)?

b. Too low (i.e., do not press sufficiently against the screen)?

12. Describe the screen-cleaning system that does not use brushes.

13. When should a screen be discarded and replaced?

14. Measure the dimensions of a wire mesh screen and a perforated sheet metal screen, and answer the following, using technical data from the manufacturer's publications:

    Cleaner

    Screen length and width

    Length of blank apron

    a. Wire mesh screen:

        Opening size

        Total square area

        Area of apron

        Area of ribs and frame

        Area containing openings

        % Useful area (i.e., with openings)

b. Perforated sheet metal screen:

% Open space

Perforation size

Total screen area

Area of apron

Area of ribs and frame

% Useful area (i.e., with perforations)

% Open space

15. Place the following screen sizes in order of increasing size of perforation (i.e., beginning with the smallest perforation):

a. Round perforations: 19, 12, 1/25, 1/14, 6, 23, 1/19, 1/15, 11

b. Oblong perforations: 7/64 × 3/4, 1/16 × 1/2, 9/64 × 3/4, 1/20 × 1/2, 1/13 × 3/4

c. Wire mesh: 6 × 23, 6 × 22, 6 × 26, 6 × 18

16. List the two types of separations made by screens and describe the type of materials removed from seed by each type of screen.

17. a. On a scalping screen, the good seed ________________ the screen perforations.

    b. On a grading screen, the good seed ________________ the screen perforations.

18. Special separations according to specific aspects of seed size and/or shape are also made by screens; for example,

    a. To separate seed by width in the air-screen cleaner, use a screen with ________________ perforations.

    b. To separate seed by thickness in the air-screen cleaner, use a screen with ________________ perforations.

19. Listed below are some screen sizes. For each, give the shape of its perforation and the kind of material from which the screen surface is made.

    a. 17

    b. 6 × 3/4

    c. 1/22

d. 9 Tri

e. 1/15 × 1/2

f. 14 × 14

g. 6 × 24

20. Points, conditions, or elements of special interest.

chapter forty

Screen selection

Conditioner name:

Date:

Seed conditioning plant/machine location:

Machine brand, model number, other identification:

Year machine installed; condition:

Installation and feeding seed/waste materials to/from the machine description:

Any problems encountered with operation of the machine:

Procedure: Use samples, hand screens, screens, and air-screen cleaner as required. Provide the following information or answers to questions.

1. Why are screens with different perforation sizes/shapes used?

2. Screens are selected, and installed in the air-screen cleaner, to make two kinds of separations. Name and describe them.

3. Scalping screen perforations are selected to obtain the ________________.

4. Grading screen perforations are selected to obtain the ________________.

5. For round seed, screen perforation shapes usually are

   a. Scalping screens ________________________________

   b. Grading screens ________________________________

6. For elongated seed, screen perforation shapes usually are

   a. Scalping screens ________________________________

   b. Grading screens ________________________________

7. For lens-shaped seed, screen perforation shapes usually are

   a. Scalping screens ________________________________

   b. Grading screens ________________________________

8. Why does the "common" seed air-screen cleaner use four consecutive screens—two scalping and two grading—instead of only two screens (i.e., one of each kind)?

9. Using two consecutive scalping screens, which perforation size is used first? Why?

10. Using two consecutive grading screens, which perforation size is used first? Why?

11. How are approximate screen perforation sizes determined before the seed lot is cleaned?

12. Can exact screen perforation sizes be determined with hand screens? Is it sometimes necessary to change a preselected screen when actual cleaning begins? Why?

13. Use hand screens to determine for each of several samples:

    a. Column 1: the smallest perforation size that permits all good seed to pass through the perforation (scalping screen).
    b. Column 2: the largest perforation size that retains the god seed, that is, does not permit more than a few good seed to pass through (grading screen).

Sample no.	Crop and variety	Type of perforation	Column 1 (scalping)	Column 2 (grading)
1		Round		
		Oblong		
2		Round		
		Oblong		
3		Round		
		Oblong		
4		Round		
		Oblong		
5		Round		
		Oblong		
6		Round		
		Oblong		
7		Round		
		Oblong		
8		Round		
		Oblong		

14. Use hand screens to determine the correct screen perforation sizes for a four-screen air-screen cleaner for the above samples. Use two scalping and two grading screens for each. List screen sizes selected and briefly explain why they were selected.

Sample no.	Crop and variety	Type of separation	First screen	Second screen	Why selected
1		Scalp			
		Grade			
2		Scalp			
		Grade			
3		Scalp			
		Grade			
4		Scalp			
		Grade			
5		Scalp			
		Grade			
6		Scalp			
		Grade			
7		Scalp			
		Grade			
8		Scalp			
		Grade			

15. Split soybeans can be removed, fairly well, by the air-screen cleaner, by using a screen with ____________ perforations as a ____________ screen.

16. Given screens ____________

Arrange them in the proper order of use in the air-screen cleaner, using screens no. 1 and 3 as scalping screens.

(1) ____________ (2) ____________

(3) ____________ (4) ____________

17. Use hand screens to demonstrate and study the following concept:
    Sometimes an elongated seed (e.g., a small grain seed) will excessively plug slotted perforations in a grading screen, so an adequate separation cannot be made. In such cases, what shape of perforation can be substituted? Why?

18. When a grading screen is used to separate a specific contaminant (as contrasted to general separation of all smaller materials), it is usually used as the second grading screen. Why?

19. To make a complete screen separation, each seed must have one or more opportunities to ____.

20. Points, conditions, or elements of special interest.

chapter forty-one

Air-screen cleaner operation

Conditioner name:

Date:

Seed conditioning plant/machine location:

Machine brand, model number, other identification:

Year machine installed; condition:

Installation and feeding seed/waste materials to/from the machine description:

Any problems encountered with operation of the machine:

Procedure: Set up, adjust, and operate the air-screen cleaner on a seed lot as described. Provide the following information or answers to questions.

1. Seed crop and variety:

   Uncleaned (original):

   Weight

   Test weight

   % Germination

   % Purity and purity analysis (identify unwanted seed)

2. Purpose of using the air-screen cleaner.

   a. Is a maintenance manual available, or an operator's manual with a detailed description of recommended maintenance? Is the machine in good condition and well maintained? What changes/improvements in maintenance would improve it?

3. Conditioning previously done to this seed lot.

4. Air-screen cleaner make and model.

   Number of screens for each kind of separation

   Number and location of air separations

5. Adjustments, screens, etc., used.

Adjustment	Setting, screen size, etc., used

6. Diagram the flow of good seed through the cleaner.

7. Which screens—positions in the cleaner and perforation sizes—are used to remove material larger than the good seed?

8. Which screens—positions in the cleaner and perforation sizes—are used to remove material smaller than the good seed?

9. With the air-screen cleaner and screens used this way, what are the positions of

   a. The adjustable gate in the pan beneath screen no. 2 (if in this model)?

   b. The adjustable gate at the discharge end of screen no. 2 (if in this model)?

10. Cleaning/separating results.

Spout	Material	% Pure seed	Test weight	% Germination	% of original lot weight

11. Was the desired separation accomplished? What further conditioning is required?

12. Were many good seed lost with the separated waste materials? How could this loss be reduced? How can lost good seed be recovered?

13. Make significant changes in adjustments as indicated below, and note the results. (After an adjustment is noted, return it to normal operation before changing the next adjustment.)

    a. Increase feed rate

    b. Decrease feed rate

c. Increase screen vibration

d. Decrease screen vibration

e. Increase upper air

f. Decrease upper air

g. Increase lower air

h. Decrease lower air

i. Connect screen tappers (operating)

j. Disconnect screen tappers (not operating)

k. Increase (steepen) pitch of scalper screens

l. Decrease (flatten) pitch of scalper screens

m. Increase (steepen) pitch of grading screens

n. Decrease (flatten) pitch of grading screens

o. Open gate beneath screen no. 2 (if in this model)

p. Close gate beneath screen no. 2 (if in this model)

q. Open gate at end of screen no. 2 (if in this model)

r. Close gate at end of screen no. 2 (if in this model).

s. Exchange the two scalper screens

t. Exchange the two grading screens

u. Exchange scalper screen no. 1 and grader screen no. 1

v. Exchange scalper screen no. 2 and grader screen no. 2

14. Adjust feed roll to a medium speed, and adjust the feed gate above the feed roll to give a uniform flow of seed at approximately the correct rate. Then, what happens to the rate of feed when the speed of the feed roll is

    a. Increased?

    b. Decreased?

15. From the correct feed rate, adjust feed gate above the feed roll as indicated below, and note the results.

    a. Raise the gate

    b. Lower the gate

16. What happens when the outside gate below the feed roll is

    a. Opened?

    b. Closed?

17. Points, conditions, or elements of special interest.

chapter forty-two

Cylinder separator characteristics

Conditioner name:

Date:

Seed conditioning plant/machine location:

Machine brand, model number, other identification:

Year machine installed; condition:

Installation and feeding seed/waste materials to/from the machine description:

Any problems encountered with operation of the machine:

Procedure: Study the cylinder separator and provide the following information or answers to questions.

1. Describe the seed separating principle(s) used and the seed characteristics that affect the separation.

2. Describe how the separation is made.

3. On what crop seed is the cylinder separator used? Why?

4. From where does the crop seed come as it goes to the cylinder separator? Why?

5. Where does the crop seed go as it leaves the cylinder separator? Why?

6. Is an operator's manual available? What is the machine designed to do? Is it being used for its designed purpose? If not, how is it being used? Can its use be changed to take better advantage of its design features? Is another machine(s) needed to improve cleaning performance?

7. Where is seed handled by this machine coming from, and where is it being fed to?

8. What additional facilities, structures, subsidiary equipment, etc., are required for this machine to operate effectively and efficiently?

Component, etc.	Needed for this machine	Comments
a. Elevator bringing seed not previously conditioned on this machine to the holding bin serving this machine. This elevator should have a two-way valve on its discharge, so that seed may bypass this machine if it is not needed.	Needed	To fit the machine's installation
b. Spouting of seed (coming from the previous separator) to this elevator.	Needed	To fit the machine's installation
c. Holding bin mounted over the feed intake of this machine.	Needed	To fit the machine's installation
d. Spouting of seed from this elevator to the holding bin mounted over this machine.	Needed	To fit the machine's installation
e. Spouting or flow of seed from the bin into the feed intake of this machine.	Needed	To fit the machine's installation
f. Mounting frame to support this machine in a position and at a height required for efficient operation.	Needed	To fit the machine's installation
g. Worker platforms, with required access steps or ladders, in appropriate locations so that operators can easily reach all controls of the machine for adjustments, and all parts of the machine for clean-out, setup, and repair.	Needed	To fit the machine's installation
h. Spouting or flow system to take good seed leaving this machine to the bin or conveyor/elevator taking good seed to the next machine in the conditioning sequence.	Needed	To fit the machine's installation
i. Dust and dusty air exhaust ducting and outside collector(s) handling dusty air and/or light materials separated from the good seed by this machine.	As needed	If needed, to fit the machine's installation
j. If the air around this machine is dusty, an exhaust fan/ducting/collection system should be installed to remove all dusty air.	As needed	If needed, to fit the machine's installation
k. Spouting to take each fraction of removed undesirable materials either to (1) bag collecting this material at this conditioning machine or to (2) a conveyor system taking all fractions of separated undesirable materials to a central waste product collecting system.	Needed	To fit the machine's installation
l. Elevator and required spouting to collect middling fraction(s) discharged from this machine back into the good-seed bin serving this machine, for reconditioning.	Not usually needed	Not needed

9. Describe the forces that first hold seed in the indents and later cause them to fall out.

10. What are the usual uses of the cylinder separator in seed conditioning? What other uses are possible?

11. What is the usual position of the cylinder separator in the seed conditioning line or sequence?

12. What conditioning is usually done to a seed lot before it goes to the cylinder separator? Why?

13. Make and model.

14. Name and address of manufacturer.

15. Draw a simple diagram to illustrate seed flow into and through the cylinder separator, and how all separated fractions discharge from the separator. Label all seed fractions and essential parts of the separator.

    a. Side view

    b. End view

16. a. Number of cylinders in the separator.

Cylinder length and diameter.

b. Type of liftings trough.

17. What are the cylinder separator's requirements for

a. Power (including RPM)?

b. Installation and mounting?

c. Seed supply and feed?

d. Handling discharged good seed?

e. Handling discharged waste products?

18. Overall dimensions:

    Height

    Width

    Length

    Height, floor to seed feed intake

    Height, floor to good seed discharge

19. Draw the separator's mounting base (i.e., the part that sits on the floor). Show all dimensions, as well as the position of the seed feed intake, good seed discharge spout, and all waste product discharge spouts.

20. List and describe all adjustments, changes, etc., that can be made to affect the separation.

21. How is seed feed or intake adjusted? Can seed feed be stopped while the cylinder is still running? How?

22. Can the speed of rotation of the cylinder be adjusted? How? Why?

23. Can the end slope or longitudinal inclination of the cylinder be adjusted? How? Why?

24. Can the cylinder(s) be changed? Why? How? How much time is required?

25. Describe how the level of the seed mass inside the cylinder should be maintained.

26. How is the level of the seed mass in the cylinder controlled?

27. How and where is the short seed fraction discharged from the separator?

28. How and where is the long seed fraction discharged from the separator?

29. a. Can the position of the separating or "leading" edge of the liftings trough be adjusted?

    b. How?

    c. Why?

30. How is the adjustable retarder in the long (not lifted) seed discharge adjusted? How does it affect the separation?

31. How can you check the separation being made?

32. Cylinder rotation speed too high results in ______________________________

33. Cylinder rotation speed too low results in ______________________________.

34. Leading edge of the liftings trough set too high results in ____________________.

35. Leading edge of the liftings trough set too low results in ____________________.

36. What would cause some long seed to be lifted up into the liftings trough, so they discharge with the short seed?

    a.

    b.

    c.

chapter forty-three

Cylinders

Conditioner name:

Date:

Seed conditioning plant/machine location:

Machine brand, model number, other identification:

Year machine installed; condition:

Installation and feeding seed/waste materials to/from the machine description:

Any problems encountered with operation of the machine:

Procedure: Study cylinders, indent sizes, and their selection. Provide the following information or answers to questions.

1. From where does the crop seed come as it goes to the cylinder separator? Why?

2. Where does the crop seed go as it leaves the air-screen cleaner?

3. Can the cylinder(s) in the separator be changed? How? Why?

4. Describe the corrective adjustments made (instead of changing the cylinder) when the installed cylinder indent size is slightly larger than the ideal.

5. Describe the corrective adjustments made (instead of changing the cylinder) when the installed cylinder indent size is slightly smaller than the ideal.

6. How many separated fractions are produced by one cylinder? Describe.

7. How many indent sizes are in one cylinder? Why?

8. How can more separated fractions be produced with the cylinder separator?

9. How can capacity of the length separation be increased?

10. How are indentations (pockets) in the cylinders measured and identified?

11. List cylinder sizes available and their primary uses.

Indent no.	Will lift	Will reject

12. How is the correct indent size determined?

13. In a seed conditioning plant that has a high volume and often needs different cylinder indent sizes, what can be done instead of constantly changing cylinders?

14. Use the laboratory model cylinder separator to determine the correct cylinder indent size to make the desired separation on several seed samples.

Sample no.	Crop seed	Contaminant to remove	Cylinder indent size to use

15. Points, conditions, or elements of special interest.

chapter forty-four

Cylinder separator operation

Conditioner name:

Date:

Seed conditioning plant/machine location:

Machine brand, model number, other identification:

Year machine installed; condition:

Installation and feeding seed/waste materials to/from the machine description:

Any problems encountered with operation of the machine:

Procedure: Using a seed lot, set up, adjust, and operate the cylinder separator as described. Provide the following information or answers to questions.

1. Seed crop and variety

   % Purity

   Test weight

   Previous conditioning done to this seed lot

   Impurities to be removed

2. Make and model of cylinder separator.

3. Is a maintenance manual available, or an operator's manual with a detailed description of recommended maintenance? Is the machine in good condition and well maintained? What changes/improvements in maintenance would improve it?

4. Size of cylinder indentations.

5. Diagram the comparative length, etc., of the crop seed and the impurities to be removed.

6. Approximate adjustments used.

Adjustment	Setting used

7. Separating results.

Spout	Material discharged	Remarks

8. a. Was the desired separation accomplished? Does the seed lot now meet standards? What further conditioning is required?

   b. Was much good seed lost with the separated waste products? How could this loss be prevented or reduced? How can the lost good seed be recovered?

9. Make the following changes in adjustments, and note the results. (After noting one adjustment, return it to normal operation before changing the next adjustment.)

   a. Increase cylinder rotation speed.

   b. Decrease cylinder rotation speed.

   c. Lower the leading (separating) edge of the liftings trough.

   d. Raise the leading (separating) edge of the liftings trough.

   e. Increase rate of seed feed.

f. Decrease rate of seed feed.

g. Raise adjustable gate in discharge spout for long seed.

10. Diagram the flow of each fraction separated by the cylinder separator.

11. Points, conditions, or elements of special interest.

chapter forty-five

Disc separator characteristics

Conditioner name:

Date:

Seed conditioning plant/machine location:

Machine brand, model number, other identification:

Year machine installed; condition:

Installation and feeding seed/waste materials to/from the machine description:

Any problems encountered with operation of the machine:

Procedure: Study the disc separator and provide the following information or answers to questions.

1. a. Seed separating principle(s) used.

   b. Illustrate this principle with simple drawings.

2. Describe how the separation is made.

3. On what crop seed is the disk separator used? Why?

4. From where does the crop seed come as it goes to the disk separator? Why?

5. Where does the crop seed go as it leaves the disk separator?

6. Is an operator's manual available? What is the machine designed to do? Is it being used for its designed purpose? If not, how is it being used? Can its use be changed to take better advantage of its design features? Is another machine(s) needed to improve cleaning performance?

7. What are the usual uses of the disc separator in seed conditioning? What other uses are possible?

Component, etc.	Needed for this machine	Comments
a. Elevator bringing seed not previously conditioned on this machine to the holding bin serving this machine. This elevator should have a two-way valve on its discharge, so that seed may bypass this machine if it is not needed.	Needed	To fit the machine's Installation
b. Spouting of seed (coming from the previous separator) to this elevator.	Needed	To fit the machine's installation
c. Holding bin mounted over the feed intake of this machine.	Needed	To fit the machine's installation
d. Spouting of seed from this elevator to the holding bin mounted over this machine.	Needed	To fit the machine's installation
e. Spouting or flow of seed from the bin into the feed intake of this machine.	Needed	To fit the machine's installation
f. Mounting frame to support this machine in a position and at a height required for efficient operation.	Needed	To fit the machine's installation
g. Worker platforms, with required access steps or ladders, in appropriate locations so that operators can easily reach all controls of the machine for adjustments, and all parts of the machine for clean-out, setup, and repair.	Needed	To fit the machine's installation
h. Spouting or flow system to take good seed leaving this machine to the bin or conveyor/elevator taking good seed to the next machine in the conditioning sequence.	Needed	To fit the machine's installation
i. Dust and dusty air exhaust ducting and outside collector(s) handling dusty air and/or light materials separated from the good seed by this machine.	As needed	If needed, to fit the machine's installation
j. If the air around this machine is dusty, an exhaust fan/ducting/collection system should be installed to remove all dusty air.	As needed	If needed, to fit the machine's installation
k. Spouting to take each fraction of removed undesirable materials either to (1) bag collecting this material at this conditioning machine or to (2) a conveyor system taking all fractions of separated undesirable materials to a central waste product collecting system.	Needed	To fit the machine's installation
l. Elevator and required spouting to collect middling fraction(s) discharged from this machine back into the good-seed bin serving this machine, for reconditioning.	Not usually needed	Not needed

8. What is the usual position of the disc separator in the seed conditioning line or sequence?

9. Where is seed handled by this machine coming from, and where is it being fed to?

10. What additional facilities, structures, subsidiary equipment, etc., are required for this machine to operate effectively and efficiently?

11. What conditioning is usually done to a seed lot before it goes to the disc separator? Why?

12. Make and model.

13. Name and address of manufacturer.

14. Draw a simple diagram to illustrate seed flow into and through the disc separator as a separation is made and how all separated fractions discharge from the separator. Label all seed fractions and essential parts of the separator.

15. What are the disc separator's requirements for

    a. Power (including RPM)?

    b. Seed supply and feed?

    c. Installation and mounting?

    d. Handling discharged good seed?

    e. Handling discharged waste products?

    f. Dust control?

16. Overall dimensions:

    Height

    Width

    Length

Height, floor to seed feed intake

Height, floor to good seed discharge

17. Draw the mounting base (i.e., the part that sits on the floor) of the disc separator. Show all dimensions, and the positions of seed feed intake, good seed discharge, and all waste discharge spouts.

18. List and describe all adjustments, changes, etc., that can be made to affect the separation.

19. a. How many discs does the separator have?

    b. What is the disc diameter?

20. Do all discs in the machine have "pockets" of the same size and shape? Describe.

21. In which end of the disc separator are the discs whose pockets are the largest? Why is it here?

22. How is the seed mass forced to flow through the open centers of the discs?

23. How can the speed of seed flow through the open centers of the discs be slowed?

24. Can the speed of disc rotation be changed? Describe.

25. How is the seed feed or intake adjusted?

26. As seed move through the disc separator, where are shorter seed and materials discharged from the separator? How?

27. Describe the function and use of the liftings return auger.

28. What is the purpose of the row of small gates on the liftings discharge side of the disc separator? How do they function? How are they adjusted? Why?

29. What is the purpose of the dividers in the liftings discharge spouts?

30. How can the retarder gate in the long-seed discharge spout be adjusted? What are its effects on the seed mass and the separation made?

31. What are grain level control blades? Where are they installed? How and why are they used?

32. How many different kinds and sizes of disc pockets are installed on the single shaft in this disc separator? How many separated fractions are produced?

33. Describe how the disc separator is cleaned out when changing varieties.

34. Describe the sequence of initial adjustments for setting up and operating the disc separator.

35. Discuss other makes, models, and kinds of disc separators available.

36. Points, conditions, or elements of special interest.

37. In the notebook that carries your study manual and your own observations, attach a copy of the operation and maintenance manuals of this model of machine that is installed in your facility.

38. Identify and describe five factors or conditions that influence the performance of this machine.

chapter forty-six

Discs

Conditioner name:

Date:

Seed conditioning plant/machine location:

Machine brand, model number, other identification:

Year machine installed; condition:

Installation and feeding seed/waste materials to/from the machine description:

Any problems encountered with operation of the machine:

Procedure: Study disc pocket sizes, types, and their selection for specific separations. Provide the following information or answers to questions.

1. Draw a disc and identify its important parts.

2. A disc separator is basically a series of discs mounted on a shaft. In a usual operation, do all discs have pockets of the same size? Describe.

3. What are the different types of disc pockets? How are they measured and identified?

   a.

   b.

   c.

4. Describe the general use of each type of disc pocket.

a.

b.

c.

5. Where are the smallest, and the largest, disc pockets located in a separator? Why?

6. List the disc pocket sizes available and their uses.

Pocket size	Will lift	Will reject

7. How is the correct disc pocket size determined?

8. What other adjustments can be made on the disc separator to correct for disc pockets slightly different from the ideal size? Discuss.

9. How can discs be changed on the disc separator? How much time is required? What is a "splitter" disc section? How is it used?

10. Using the lab model disc separator, determine the correct disc pocket shape(s) and size(s) to make the desired separation(s) on several samples.

Sample no.	Crop seed	Contaminants to remove	Disc pocket size(s) to use

11. Points, conditions, or elements of special interest.

chapter forty-seven

Disc separator operation

Conditioner name:

Date:

Seed conditioning plant/machine location:

Machine brand, model number, other identification:

Year machine installed; condition:

Installation and feeding seed/waste materials to/from the machine description:

Any problems encountered with operation of the machine:

Procedure: Using a seed lot, set up, adjust, and operate the disc separator as described. Provide the following information or answers to questions.

1. Seed crop and variety.

   Original seed lot:

   Weight

   % Germination

   % Purity

   Test weight

   Impurities

2. Purpose of using the disc separator.

3. Is a maintenance manual available, or an operator's manual with a detailed description of recommended maintenance? Is the machine in good condition and well maintained? What changes/improvements in maintenance would improve it?

4. Conditioning previously done on this seed lot.

5. Make and model of disc separator.

Number of discs

Disc diameter

6. Diagram the discs installed in the separator, showing the number of discs with each size of disc pocket. Label all disc pocket sizes, sections of the separator, feed intake, and discharge spouts.

7. Approximate adjustments used.

Adjustment	Setting used

8. Separation results.

Spout	Material discharged	Test weight	% Purity	% Germination	Weight	% of original lot weight

9. Was the desired separation accomplished? Does this seed lot now meet standards? What further conditioning is required?

10. Make significant changes in adjustments as indicated below, and note the results. (After noting one adjustment, return it to normal operation before changing the next adjustment.)

    a. Increase feed rate.

    b. Decrease feed rate.

    c. Lower retarder gate in long-seed discharge spout.

    d. Raise retarder gate in long-seed discharge spout.

e. Move liftings discharge gates, over the liftings return auger, outward.

f. Move liftings discharge gates, over the liftings return auger, inward.

g. Raise grain level control blades.

h. Lower grain level control blades.

11. Was much good seed lost with the undesirable material separated? How could this loss be reduced? How could the lost good seed be recovered?

12. The shortest seed is removed from the seed mass by discs in what section of the separator? By how many discs?

13. In what part of the separator are the longest seeds discharged? How do they arrive at the discharge spout?

14. Using a disc separator to separate a mixture of rice seed, short cross-broken rice seed, and morning glory seed, what discharges from the spout at the discharge end of the disc separator?

15. What discharges from the side spout nearest the feed end of the separator?

16. What discharges from the side spout furthest from the feed end?

17. Points, conditions, or elements of special interest.

chapter forty-eight

Gravity separator characteristics

Conditioner name:

Date:

Seed conditioning plant/machine location:

Machine brand, model number, other identification:

Year machine installed; condition:

Installation and feeding seed/waste materials to/from the machine description:

Any problems encountered with operation of the machine:

Procedure: Study the gravity separator and provide the following information or answers to questions.

1. The gravity separator uses difference in ____________________ of seed to make a separation. Describe its separating action.

2. On what crop seed is the gravity separator used? Why?

3. From where does the crop seed come as it goes to the gravity separator? Why?

4. Where does the crop seed go as it leaves the gravity separator?

5. Describe the separation made by the gravity separator on a mixture of seed that differ in

   a. Size, but have the same specific gravity.

   b. Specific gravity, but are of the same size.

   c. Both size and specific gravity.

d. Physical characteristics other than specific gravity and size.

6. Is an operator's manual available? What is the machine designed to do? Is it being used for its designed purpose? If not, how is it being used? Can its use be changed to take better advantage of its design features? Is another machine(s) needed to improve cleaning performance?

7. List and describe the separated fractions produced by the gravity separator.

8. How is the gravity separator used in seed conditioning? What are its primary uses? Are other uses possible?

9. Where is seed handled by this machine coming from, and where is it being fed to?

10. What additional facilities, structures, subsidiary equipment, etc., are required for this machine to operate effectively and efficiently?

11. What conditioning is done to seed before it goes to the gravity separator? Why?

12. Where is the gravity separator installed and used in the seed conditioning flow sequence?

13. Make and model of gravity separator used.

14. Name and address of manufacturer.

15. Overall dimensions:

    Height

    Width

    Length

    Height, floor to seed feed intake

    Height, floor to good seed discharge

16. What are the gravity separator's requirements for

    a. Power (including RPM)?

    b. Seed supply and feed?

    c. Installation and mounting?

    d. Handling discharged clean seed?

    e. Handling discharged waste fractions?

    f. Air supply to the separator?

    g. Dust control?

    h. Preventing false vibrations of the separator?

Component, etc.	Needed for this machine	Comments
a. Elevator bringing seed not previously conditioned on this machine to the holding bin serving this machine. This elevator should have a two-way valve on its discharge, so that seed may bypass this machine if it is not needed.	Needed	To fit the machine's installation
b. Spouting of seed (coming from the previous separator) to this elevator.	Needed	To fit the machine's installation
c. Holding bin mounted over the feed intake of this machine.	Needed	To fit the machine's installation
d. Spouting of seed from this elevator to the holding bin mounted over this machine.	Needed	To fit the machine's installation
e. Spouting or flow of seed from the bin into the feed intake of this machine.	Needed	To fit the machine's installation
f. Mounting frame to support this machine in a position and at a height required for efficient operation.	Needed	To fit the machine's installation
g. Worker platforms, with required access steps or ladders, in appropriate locations so that operators can easily reach all controls of the machine for adjustments, and all parts of the machine for clean-out, setup, and repair.	Needed	To fit the machine's installation
h. Spouting or flow system to take good seed leaving this machine to the bin or conveyor/elevator taking good seed to the next machine in the conditioning sequence.	Needed	To fit the machine's installation
i. Dust and dusty air exhaust ducting and outside collector(s) handling dusty air and/or light materials separated from the good seed by this machine.	As needed	If needed, to fit the machine's installation
j. If the air around this machine is dusty, an exhaust fan/ducting/collection system should be installed to remove all dusty air.	As needed	If needed, to fit the machine's installation
k. Spouting to take each fraction of removed undesirable materials either to (1) bag collecting this material at this conditioning machine or to (2) a conveyor system taking all fractions of separated undesirable materials to a central waste product collecting system.	Needed	To fit the machine's installation
l. Elevator and required spouting to collect middling fraction(s) discharged from this machine back into the good-seed bin serving this machine, for reconditioning.	Not usually needed	Not needed

17. Draw the gravity separator's mounting base (i.e., the part that sits on the floor). Show all dimensions, location of "bolt-down" points, position of seed feed intake, good seed discharge spout, and all waste product spouts.

18. List and describe all adjustments on the gravity separator.

19. Why is a uniform feed rate necessary? How is the feed rate used as an adjustment to control the separating action?

20. How is the feed rate controlled?

21. Draw a plan view of the gravity separator deck. Show all dimensions and the flow of all separated fractions.

22. The two main types of deck cover used for the gravity separator are

(1)

(2)

Illustrate by drawing each type.

23. A deck cover of ______________________ would be used for small seed, such as clovers.

24. A deck cover of ______________________ would be used for large seed, such as corn (maize).

25. What is the purpose of the large ½-inch mesh wire above the wire mesh deck cover?

26. What is the purpose of the rows of sheet metal strips on top of the wire mesh deck cover?

27. a. How is the deck fastened onto the air chest?

 b. Why must the deck be fastened securely in place?

28. How are undesirable false vibrations of the overall machine minimized?

29. The air flow is used to ______________________________________.

30. Air adjustment is proper when ________________________________.

31. How is the required flow of air created? Describe the system and its operation.

32. Draw a simple diagram to illustrate the flow of air into and through the gravity separator. Label all essential parts of the gravity.

33. How many fans does the gravity have?

    Type of fans.

    Fan diameter, number of blades, blade width, and RPM.

    Are all fans mounted on the same shaft? Describe.

34. How is the air adjusted and controlled? Describe how the controls work.

35. a. How is uniform (i.e., uniform pressure without turbulence) air distribution achieved in the air chest beneath the deck?

    b. Draw a simple diagram illustrating this uniform air distribution.

36. a. How is the deck constructed to achieve the desired air flow through all parts of the seed mass moving across the deck?

    b. Illustrate this with a simple drawing.

37. Describe the air filters used and their location.

38. Why should air filters always be in their proper positions when the gravity separator is operating?

39. How are air filters cleaned?

40. Why is the deck tilted to one side (side slope)?

41. How is the deck side slope adjusted? How much can it be changed?

42. Why is the deck tilted toward the discharge end (end slope)?

43. How is deck end slope adjusted? How much can it be changed?

44. a. Describe the system that causes the deck to oscillate and how it works.

b. Illustrate the pattern of oscillation or movement of the deck.

45. How can speed of deck oscillation be changed?

46. How does speed of deck oscillation affect the separation? Describe the use of deck oscillation speed as an adjustment.

47. The gravity separator is properly adjusted when ________________________________.

48. To move the seed mass "uphill," one or a combination of these adjustments may be used.

49. To move the seed mass "downhill," one or a combination of these adjustments may be used.

50. List and describe the "separating" adjustments.

51. List and describe the "capacity" adjustments.

52. a. How should adjustments be made?

    b. Why are adjustments made in this way?

53. At the discharge end of the deck, how are the seed fractions separated and guided to the proper discharge spouts?

54. a. How many discharge spouts are provided?

    b. Draw a diagram illustrating the discharge spouts and indicate which separated seed fraction goes to each spout.

55. How can seed fractions be handled as they leave the discharge spouts? Describe.

56. What is the "middling product?" How is it formed?

57. How can good seed in the middling product be recovered? Describe the various methods possible.

58. Diagram the various methods used to recover good seed from the middling product.

59. Describe the sequence of initial adjustments for setting up and operating the gravity separator.

60. Briefly discuss and describe other makes, models, and kinds of gravity separator available.

61. Points, conditions, or elements of special interest.

62. In the notebook that carries your study manual and your own observations, attach a copy of the operation and maintenance manuals of this model of machine that is installed in your facility.

63. Identify and describe five factors or conditions that influence the performance of this machine.

chapter forty-nine

Gravity separator operation

Conditioner name:

Date:

Seed conditioning plant/machine location:

Machine brand, model number, other identification:

Year machine installed; condition:

Installation and feeding seed/waste materials to/from the machine description:

Any problems encountered with operation of the machine:

Procedure: Using a seed lot, set up, adjust, and operate the gravity as described. Provide the following information or answers to questions.

1. Seed crop and variety.

   Test weight

   % Purity

   % Germination

   Impurities

   Seed lot weight

   Previous conditioning done on this seed slot

   Purpose of using the gravity separator

2. Gravity separator make and model.

3. Is a maintenance manual available, or an operator's manual with a detailed description of recommended maintenance? Is the machine in good condition and well maintained? What changes/improvements in maintenance would improve it?

4. Approximate adjustments used.

Adjustment	Setting used

5. Separation and results.

Fraction	Type of contaminant	Test weight	% Germination	% Purity	Weight	% of original lot

6. With the machine properly adjusted, leave other adjustments unchanged, but make significant changes, as indicated in individual adjustments and note the results. (After noting each, bring it back to normal operation before making the next change.)

   a. Decrease feed rate

   b. Increase feed rate

   c. Increase air

   d. Decrease air

   e. Increase deck oscillation speed

   f. Decrease deck oscillation speed

   g. Increase end slope

   h. Decrease end slope

i. Increase side slope

j. Decrease side slope

7. What adjustments and/or other actions would be required to correct the following conditions?

   a. Lower side of the deck has a "blank" or uncovered space.

   b. Upper side of the deck has a blank space.

   c. The "bed" of seed seems to "lie dead" on the deck and separation is poor.

   d. The seed bed "bubbles" and separation is poor.

   e. Deck shows "dead spots."

   f. Seed bed "surges" on the deck.

8. Points, conditions, or elements of special interest.

chapter fifty

Stoner characteristics

Conditioner name:

Date:

Seed conditioning plant/machine location:

Machine brand, model number, other identification:

Year machine installed; condition:

Installation and feeding seed/waste materials to/from the machine description:

Any problems encountered with operation of the machine:

Procedure: Study the stoner and provide the following information or answers to questions.

1. Seed separation principle(s) used.

2. What are the usual uses of the stoner in seed conditioning? Are other uses possible?

3. On what crop seed is the stoner used? Why?

4. From where does the crop seed come as it goes to the stoner? Why?

5. Where does the crop seed go as it leaves the stoner?

6. Is an operator's manual available? What is the machine designed to do? Is it being used for its designed purpose? If not, how is it being used? Can its use be changed to take better advantage of its design features? Is another machine(s) needed to improve cleaning performance?

7. What is the usual position of the stoner in the seed conditioning line or sequence?

   a. Where is seed handled by this machine coming from, and where is it being fed to?

b. What additional facilities, structures, subsidiary equipment, etc., are required for this machine to operate effectively and efficiently?

Component, etc.	Needed for this machine	Comments
a. Elevator bringing seed not previously conditioned on this machine to the holding bin serving this machine. This elevator should have a two-way valve on its discharge, so that seed may bypass this machine if it is not needed.	Needed	To fit the machine's installation
b. Spouting of seed (coming from the previous separator) to this elevator.	Needed	To fit the machine's installation
c. Holding bin mounted over the feed intake of this machine.	Needed	To fit the machine's installation
d. Spouting of seed from this elevator to the holding bin mounted over this machine.	Needed	To fit the machine's installation
e. Spouting or flow of seed from the bin into the feed intake of this machine.	Needed	To fit the machine's installation
f. Mounting frame to support this machine in a position and at a height required for efficient operation.	Needed	To fit the machine's installation
g. Worker platforms, with required access steps or ladders, in appropriate locations so that operators can easily reach all controls of the machine for adjustments, and all parts of the machine for clean-out, setup, and repair.	Needed	To fit the machine's installation
h. Spouting or flow system to take good seed leaving this machine to the bin or conveyor/elevator taking good seed to the next machine in the conditioning sequence.	Needed	To fit the machine's installation
i. Dust and dusty air exhaust ducting and outside collector(s) handling dusty air and/or light materials separated from the good seed by this machine.	As needed	If needed, to fit the machine's installation
j. If the air around this machine is dusty, an exhaust fan/ducting/collection system should be installed to remove all dusty air.	As needed	If needed, to fit the machine's installation
k. Spouting to take each fraction of removed undesirable materials either to (1) bag collecting this material at this conditioning machine or to (2) a conveyor system taking all fractions of separated undesirable materials to a central waste product collecting system.	Needed	To fit the machine's installation
l. Elevator and required spouting to collect middling fraction(s) discharged from this machine back into the good-seed bin serving this machine, for reconditioning.	Not usually needed	

8. What conditioning is usually done to a seed lot before it goes to the stoner? Why?

9. Make and model.

10. Name and address of the manufacturer.

11. Describe the stoner's separating action and operation.

12. Draw the stoner deck. Show all dimensions, as well as positions of seed feed, good seed discharge, and waste product discharge.

13. What are the stoner's requirements for

    a. Power (including RPM)?

    b. Installation and mounting?

    c. Seed supply and feed?

    d. Handling discharged clean seed?

    e. Handing discharged waste material?

    f. Air ducting and dust control?

14. Overall dimensions:

    Height

    Length

    Width

    Height, floor to seed feed intake

    Height, floor to good seed discharge

15. Draw the mounting base of the stoner (i.e., the part that sits on the floor). Show all dimensions, as well as the positions of seed feed intake, good seed discharge, waste product discharge, and piped air intake (if air is piped in from the outside).

16. List and describe all adjustments on the stoner.

17. Where is seed fed onto the stoner deck? How is feed rate controlled?

18. Why does the high end of the deck become narrow to a small discharge point? Why is a gate provided to close this narrow high-end discharge?

19. What deck surface covers are available? How is each used?

20. What deck construction is provided to ensure uniform air flow through the deck and the seed mass?

21. Draw a cross section of the deck.

22. Describe the mechanism and the action by which the deck oscillates.

23. How is deck oscillation speed adjusted? Describe the mechanism and action.

24. Deck drive shaft RPM: minimum and maximum.

25. Describe the mechanism by which the deck is inclined.

26. How is the deck secured in position after its slope (inclination) is changed? Why is this necessary?

27. Slope of the deck: minimum and maximum.

28. Describe the mechanism and control for adjusting the air.

29. Fan that supplies air:

    Type of fan

    RPM

    Diameter

    Width of blades

30. Inside dimensions of the air chest beneath the deck:

    Width

    Height

    Length

31. What mechanisms, designs, etc., are provided to ensure uniform air flow and air pressure within the air chest?

32. a. Explain how a "stoner-type" separation can be made on the deck of some models of gravity separator.

    b. Draw the deck in A above (this question), to illustrate the described separation. Label all essential parts.

33. Describe the sequence of initial adjustments for setting up and operating the stoner.

34. Briefly describe and discuss other makes, models, and kinds of stoners.

35. Points, conditions, or elements of special interest.

36. In the notebook that carries your study manual and your own observations, attach a copy of the operation and maintenance manuals of this model of machine that is installed in your facility.

37. Identify and describe five factors or conditions that influence the performance of this machine.

chapter fifty-one

Stoner operation

Conditioner name:

Date:

Seed conditioning plant/machine location:

Machine brand, model number, other identification:

Year machine installed; condition:

Installation and feeding seed/waste materials to/from the machine description:

Any problems encountered with operation of the machine:

Procedure: Using a seed lot, set up, adjust, and operate the stoner as described. Provide the following information or answers to questions.

1. Seed crop and variety.

   Original seed lot:

   Test weight

   % Germination

   % Purity

   Impurities

   Purpose of using stoner.

   Previous conditioning done to this seed lot.

2. Make, model, and manufacturer of stoner.

3. Is a maintenance manual available, or an operator's manual with a detailed description of recommended maintenance? Is the machine in good condition and well maintained? What changes/improvements in maintenance would improve it?

4. Approximate adjustments used.

Adjustment	Setting used

5. Separation results.

Fraction	Type of contaminant	Test weight	% Germination	% Purity	weight	% of original lot

6. Was the desired separation performed? Does the seed lot now meet standards? What further conditioning is required?

7. Was much good seed lost with the removed undesirable materials? How could this loss be prevented? How can these lost good seed be recovered?

8. Approximate capacity per hour.

9. Make significant changes in adjustments as indicated below, and note the results. (After making one adjustment, return it to normal operation before changing the next adjustment.)

   a. Increase feed rate

   b. Decrease feed rate

   c. Increase air

   d. Decrease air

   e. Increase deck slope

   f. Decrease deck slope

   g. Increase deck speed

   h. Remove gate at high end discharge

   i. Close gate at high end discharge.

10. Points, conditions, or elements of special interest.

chapter fifty-two

Pneumatic separator characteristics

Conditioner name:

Date:

Seed conditioning plant/machine location:

Machine brand, model number, other identification:

Year machine installed; condition:

Installation and feeding seed/waste materials to/from the machine description:

Any problems encountered with operation of the machine:

Procedure: Study the pneumatic separator and provide the following information or answers to questions.

1. Seed separating principle(s) used.

2. What is the usual use of the pneumatic separator in seed conditioning? Are other uses possible?

3. On what crop seed is the pneumatic separator used? Why?

4. From where does the crop seed come as it goes to the pneumatic separator? Why?

5. Where does the crop seed go as it leaves the pneumatic separator? Why??

6. Is an operator's manual available? What is the machine designed to do? Is it being used for its designed purpose? If not, how is it being used? Can its use be changed to take better advantage of its design features? Is another machine(s) needed to improve cleaning performance?

7. What is the usual position in the conditioning line/sequence of the pneumatic separator? What conditioning is usually done on a seed lot before it reaches the pneumatic separator?

8. Make and model.

9. Name and address of manufacturer.

10. Diagram the flow of seed and air through the pneumatic separator. Label all essential parts of the separator and label the separated fractions.

11. Draw the separating chamber of the pneumatic separator. Show dimensions and label essential parts.

   a. Where is seed handled by this machine coming from, and where is it being fed to?

   b. What additional facilities, structures, subsidiary equipment, etc., are required for this machine to operate effectively and efficiently?

12. What are the pneumatic separator's requirements for

   a. Power (including RPM)?

   b. Installation and mounting?

   c. Seed supply and feed?

d. Handling discharged good seed?

e. Handling discharged waste materials?

f. Air ducting and dust control?

13. Overall dimensions:

Height

Width

Length

Height, floor to seed feed intake

Height, floor to good seed discharge

14. Draw the mounting base (i.e., part that sits on the floor) of the pneumatic separator. Show all dimensions, position of seed intake, good seed discharge spout, waste product discharge spouts, and air exhaust duct.

15. List and describe all adjustments on the pneumatic separator.

16. Describe the seed feed intake control and how it works. Can seed feed be stopped while the separator is still running?

Component, etc.	Needed for this machine	Comments
a. Elevator bringing seed not previously conditioned on this machine to the holding bin serving this machine. This elevator should have a two-way valve on its discharge, so that seed may bypass this machine if it is not needed.	Needed	To fit the machine's installation
b. Spouting of seed (coming from the previous separator) to this elevator.	Needed	To fit the machine's installation
c. Holding bin mounted over the feed intake of this machine.	Needed	To fit the machine's installation
d. Spouting of seed from this elevator to the holding bin mounted over this machine.	Needed	To fit the machine's installation
e. Spouting or flow of seed from the bin into the feed intake of this machine.	Needed	To fit the machine's installation
f. Mounting frame to support this machine in a position and at a height required for efficient operation.	Needed	To fit the machine's installation
g. Worker platforms, with required access steps or ladders, in appropriate locations so that operators can easily reach all controls of the machine for adjustments, and all parts of the machine for clean-out, setup, and repair.	Needed	To fit the machine's installation
h. Spouting or flow system to take good seed leaving this machine to the bin or conveyor/elevator taking good seed to the next machine in the conditioning sequence.	Needed	To fit the machine's installation
i. Dust and dusty air exhaust ducting and outside collector(s) handling dusty air and/or light materials separated from the good seed by this machine.	As needed	If needed, to fit the machine's installation
j. If the air around this machine is dusty, an exhaust fan/ducting/collection system should be installed to remove all dusty air.	As needed	If needed, to fit the machine's installation
k. Spouting to take each fraction of removed undesirable materials either to (1) bag collecting this material at this conditioning machine or to (2) a conveyor system taking all fractions of separated undesirable materials to a central waste product collecting system.	Needed	To fit the machine's installation
l. Elevator and required spouting to collect middling fraction(s) discharged from this machine back into the good-seed bin serving this machine, for reconditioning.	Not usually needed	Not needed

17. Describe the air control, its location, and how it adjusts the air flow.

18. Air for the separation is provided by a fan. What is

    Type of fan

    RPM

    Fan diameter

    Width of blades

    Shaft diameter

    Number and diameter or dimensions of air intakes

    Describe fan drive

19. Describe handling of dust and exhausted air.

20. How many separated fractions does the pneumatic separator produce? Describe each.

21. a. What is the normal average capacity of the pneumatic separator?

b. Is a single pneumatic separator capable of matching the capacity of a "commercial-size" air-screen cleaner, or is more than one pneumatic separator installed together? Describe.

22. Describe the sequence of initial adjustments for setting up and operating the pneumatic separator.

23. Briefly discuss and describe other makes, models, and kinds of pneumatic separators available.

24. Points, conditions, or elements of special interest.

25. In the notebook that carries your study manual and your own observations, attach a copy of the operation and maintenance manuals of this model of machine that is installed in your facility.

26. Identify and describe five factors or conditions that influence the performance of this machine.

chapter fifty-three

Pneumatic separator operation

Conditioner name:

Date:

Seed conditioning plant/machine location:

Machine brand, model number, other identification:

Year machine installed; condition:

Installation and feeding seed/waste materials to/from the machine description:

Any problems encountered with operation of the machine:

Procedure: Using a seed lot, set up, adjust, and operate the pneumatic separator as described. Provide the following information or answers to questions.

1. Seed crop and variety.

   Original seed lot:

   Weight

   Test weight

   % Germination

   % Purity

   Impurities

   Purpose of pneumatic separation

   Previous conditioning done to this seed lot

2. Make, model, and manufacturer of the pneumatic separator.

3. Is a maintenance manual available, or an operator's manual with a detailed description of recommended maintenance? Is the machine in good condition and well maintained? What changes/improvements in maintenance would improve it?

4. Approximate adjustments used.

Adjustment	Setting used

5. Separation results.

Fraction	Type of contaminant	Test weight	% Germination	% Purity	Weight	% of original lot

6. Was the desired separation accomplished? Does this seed lot now meet standards? What further conditioning is required?

7. Make significant changes in adjustments as indicated below, and note the results. (After noting one adjustment, return it to normal operation before changing the next adjustment.)

   a. Increase feed rate

   b. Decrease feed rate

   c. Increase air

   d. Decrease air

e. Narrow the separating air column (if possible on this model)

f. Widen the separating air column (if possible on this model)

8. Were many good crop seed lost with the undesirable materials removed? How could this loss be reduced? How could the lost good seed be recovered?

9. Points, conditions, or elements of special interest.

chapter fifty-four

Aspirator characteristics

Conditioner name:

Date:

Seed conditioning plant/machine location:

Machine brand, model number, other identification:

Year machine installed; condition:

Installation and feeding seed/waste materials to/from the machine description:

Any problems encountered with operation of the machine:

Procedure: Study the aspirator and provide the following information or answers to questions.

1. Seed separating principle(s) used.

2. What is the primary use of the aspirator? What other uses does it have?

3. On what crop seed is the aspirator used? Why?

4. From where does the crop seed come as it goes to the aspirator? Why?

5. Where does the crop seed go as it leaves the aspirator?

6. Is an operator's manual available? What is the machine designed to do? Is it being used for its designed purpose? If not, how is it being used? Can its use be changed to take better advantage of its design features? Is another machine(s) needed to improve cleaning performance?

7. Should seed be partially conditioned before going to the aspirator? Why? Describe.

8. What is the usual location of the aspirator in the conditioning flow sequence?

9. Make and model.

Component, etc.	Needed for this machine	Comments
a. Elevator bringing seed not previously conditioned on this machine to the holding bin serving this machine. This elevator should have a two-way valve on its discharge, so that seed may bypass this machine if it is not needed.	Needed	To fit the machine's installation
b. Spouting of seed (coming from the previous separator) to this elevator.	Needed	To fit the machine's installation
c. Holding bin mounted over the feed intake of this machine.	Needed	To fit the machine's installation
d. Spouting of seed from this elevator to the holding bin mounted over this machine.	Needed	To fit the machine's installation
e. Spouting or flow of seed from the bin into the feed intake of this machine.	Needed	To fit the machine's installation
f. Mounting frame to support this machine in a position and at a height required for efficient operation.	Needed	To fit the machine's installation
g. Worker platforms, with required access steps or ladders, in appropriate locations so that operators can easily reach all controls of the machine for adjustments, and all parts of the machine for clean out, set-up, and repair.	Needed	To fit the machine's installation
h. Spouting or flow system to take good seed leaving this machine to the bin or conveyor/elevator taking good seed to the next machine in the conditioning sequence.	Needed	To fit the machine's installation
i. Dust and dusty air exhaust ducting and outside collector(s) handling dusty air and/or light materials separated from the good seed by this machine.	As needed	If needed, to fit the machine's installation
j. If the air around this machine is dusty, an exhaust fan/ducting/collection system should be installed to remove all dusty air.	As needed	If needed, to fit the machine's installation
k. Spouting to take each fraction of removed undesirable materials either to (1) bag collecting this material at this conditioning machine or to (2) a conveyor system taking all fractions of separated undesirable materials to a central waste product collecting system.	Needed	To fit the machine's installation
l. Elevator and required spouting to collect middling fraction(s) discharged from this machine back into the good-seed bin serving this machine, for reconditioning.	Not usually needed	Not needed

10. Name and address of manufacturer.

11. Draw a simple diagram illustrating the flow of both seed and air into, through, and out of the aspirator as the separation is made.

12. What are the aspirator's requirements for

    a. Power (including RPM)?

    b. Installation and mounting?

    c. Seed supply and feed?

    d. Handling discharged waste products?

    e. Handling discharged good seed?

    f. Air ducting and dust collection?

13. Overall dimensions:

    Height

    Width

    Length

    Height, floor to seed feed intake

    Height, floor to good seed discharge

14. Draw the mounting base (i.e., the part that sits on the floor) of the aspirator. Show all dimensions, as well as position of seed feed intake, good seed discharge, and waste product discharge spouts.

    a. Where is seed handled by this machine coming from, and where is it being fed to?

    b. What additional facilities, structures, subsidiary equipment, etc., are required for this machine to operate effectively and efficiently?

15. List and describe all adjustments on the aspirator.

16. How is the seed feed regulated? How are seed fed into the aspirator?

17. Can the seed feed be stopped while the aspirator is still running? How?

18. Is a screen or other auxiliary separating mechanism built into the aspirator? Describe.

19. How is the separating air flow controlled? Describe how the control(s) works.

20. Can the dimensions (i.e., cross-sectional area) of the air separation column be adjusted or changed? How? Why?

21. What is the RPM of the fan? Can this be adjusted or changed? How?

22. How many fractions of separated material does this aspirator produce? Describe each.

23. How are the separated fractions discharged from the aspirator? Describe.

24. Describe the sequence of initial adjustments for setting up and operating the aspirator.

25. Briefly discuss and describe other makes, models, and kinds of aspirators available.

26. Points, conditions, or elements of special interest.

27. In the notebook that carries your study manual and your own observations, attach a copy of the operation and maintenance manuals of this model of machine that is installed in your facility.

28. Identify and describe five factors or conditions that influence the performance of this machine.

chapter fifty-five

Aspirator operation

Conditioner name:

Date:

Seed conditioning plant/machine location:

Machine brand, model number, other identification:

Year machine installed; condition:

Installation and feeding seed/waste materials to/from the machine description:

Any problems encountered with operation of the machine:

Procedure: Using a suitable seed lot, set up, adjust, and operate the aspirator as described. Provide the following information or answers to questions.

1. Seed crop and variety.

   Seed lot:

   Weight

   Test weight

   % Purity

   % Germination

2. Make, model, and manufacturer of aspirator.

3. Is a maintenance manual available, or an operator's manual with a detailed description of recommended maintenance? Is the machine in good condition and well maintained? What changes/improvements in maintenance would improve it?

4. Approximate adjustments used.

Adjustment	Setting used

5. Separation results.

Spout	Type of material	Test weight	% Germination	% Purity	Weight	% of original lot

6. a. Are any undesirable materials remaining in the good seed?

b. How has quality of the seed lot been improved? Does it now meet standards?

c. What further conditioning is required?

7. Approximate capacity per hour.

8. Make significant changes in adjustments as indicated below, and note the results. (After noting one adjustment, return it to normal operation before changing the next adjustment.)

   a. Increase feed rate

   b. Decrease feed rate

   c. Increase air

   d. Decrease air

   e. Increase size (cross-sectional area) of separating air column

   f. Decrease size of separating air column

9. Points, conditions, or elements of special interest.

chapter fifty-six

Spiral separator characteristics

Conditioner name:

Date:

Seed conditioning plant/machine location:

Machine brand, model number, other identification:

Year machine installed; condition:

Installation and feeding seed/waste materials to/from the machine description:

Any problems encountered with operation of the machine:

Procedure: Study the spiral separator and provide the following information or answers to questions.

1. a. The spiral separator separates seed by differences in ______________________.

   b. Describe how the separation is made.

2. Describe the usual use of the spiral separator in seed conditioning. Are other uses possible?

3. On what crop seed is the spiral separator used? Why?

4. From where does the crop seed come as it goes to the spiral separator? Why?

5. Where does the crop seed go as it leaves the spiral separator?

6. Is an operator's manual available? What is the machine designed to do? Is it being used for its designed purpose? If not, how is it being used? Can its use be changed to take better advantage of its design features? Is another machine(s) needed to improve cleaning performance?

7. What conditioning is done to a seed lot before it goes to the spiral separator? Why?

Component, etc.	Needed for this machine	Comments
a. Elevator bringing seed not previously conditioned on this machine to the holding bin serving this machine. This elevator should have a two-way valve on its discharge, so that seed may bypass this machine if it is not needed.	Needed	To fit the machine's installation
b. Spouting of seed (coming from the previous separator) to this elevator.	Needed	To fit the machine's installation
c. Holding bin mounted over the feed intake of this machine.	Needed	To fit the machine's installation
d. Spouting of seed from this elevator to the holding bin mounted over this machine.	Needed	To fit the machine's installation
e. Spouting or flow of seed from the bin into the feed intake of this machine.	Needed	To fit the machine's installation
f. Mounting frame to support this machine in a position and at a height required for efficient operation.	Needed	To fit the machine's installation
g. Worker platforms, with required access steps or ladders, in appropriate locations so that operators can easily reach all controls of the machine for adjustments, and all parts of the machine for clean-out, setup, and repair.	Needed	To fit the machine's installation
h. Spouting or flow system to take good seed leaving this machine to the bin or conveyor/elevator taking good seed to the next machine in the conditioning sequence.	Needed	To fit the machine's installation
i. Dust and dusty air exhaust ducting and outside collector(s) handling dusty air and/or light materials separated from the good seed by this machine.	As needed	If needed, to fit the machine's installation
j. If the air around this machine is dusty, an exhaust fan/ducting/collection system should be installed to remove all dusty air.	As needed	If needed, to fit the machine's installation
k. Spouting to take each fraction of removed undesirable materials either to (1) bag collecting this material at this conditioning machine or to (2) a conveyor system taking all fractions of separated undesirable materials to a central waste product collecting system.	Needed	To fit the machine's installation
l. Elevator and required spouting to collect middling fraction(s) discharged from this machine back into the good-seed bin serving this machine, for reconditioning.	Not usually needed	Not needed

8. At what point in the conditioning flow sequence is the spiral separator usually installed and used? Why?

9. Make and model.

10. Name and address of manufacturer.

11. Overall dimensions:

    Height

    Width

    Length

    Height, floor to seed feed intake

    Height, floor to good seed discharge

12. Draw the separator's mounting base (i.e., the part that sits on the floor). Show all dimensions, position of the seed feed or intake, good seed discharge spout, and waste product spout.

13. What are the spiral separator's requirements for

    a. Installation and mounting?

    b. Seed supply and feed?

    c. Handling discharged fractions?

    d. Power?

14. Where is seed handled by this machine coming from, and where is it being fed to?

15. What additional facilities, structures, subsidiary equipment, etc., are required for this machine to operate effectively and efficiently?

16. List and describe all adjustments or controls on the spiral separator.

17. Can the rate of seed feed be adjusted? How?

18. Draw a cutaway side view showing the central column, separating spirals, and outer spiral.

    a. What is the angle (from the vertical, as measured from the vertical center column) of the sheet-metal separating spirals?

    b. How many separating spirals does the machine have?

    c. Are all separating spirals identical? Describe.

    d. Can the angle of the separating spirals be adjusted?

    e. What is the width of the separating spirals?

    f. How many complete turns around the center column do the separating spirals make?

g. What is the vertical length of the column of separating spirals?

h. What is the horizontal distance from the center column to the outer edge of the separating spirals?

19. a. How many outer spirals are there? What is its purpose?

b. What is the width of the outer spiral?

c. What is the horizontal distance from the outer edge of the outer spiral to

i. The center column?

ii. The outer edge of the separating columns?

d. What is the height of the vertical sidewall part of the outer spiral?

20. a. How many separated fractions (i.e., number of discharge spouts there are) are produced by the spiral separator?

    b. Describe the seed/particles of each of the separated fractions.

21. Is the spiral separator normally used singly or in a multiple installation? Explain. How many are required to match the usual capacity of a "commercial-size" air-screen cleaner?

22. List and describe the initial sequence of adjustments for setting up and operating the spiral separator.

    a. Briefly discuss and describe other spiral separators available.

    b. Points, conditions, or elements of special interest.

23. In the notebook that carries your study manual and your own observations, attach a copy of the operation and maintenance manuals of this model of machine that is installed in your facility.

24. Identify and describe five factors or conditions that influence the performance of this machine.

chapter fifty-seven

Spiral separator operation

Conditioner name:

Date:

Seed conditioning plant/machine location:

Machine brand, model number, other identification:

Year machine installed; condition:

Installation and feeding seed/waste materials to/from the machine description:

Any problems encountered with operation of the machine:

Procedure: Using a seed lot, set up, adjust, and operate the spiral separator as described below. Provide the following information or answers to questions.

1. Seed crop and variety.

   Seed lot:

   Weight

   % Purity

   % Germination

   Test weight

   Impurities

   Purpose of using spiral separator

   Previous conditioning done to this seed lot

2. Make, model, and manufacturer of spiral separator.

   Number of spiral separators installed as a unit.

   Describe the installation.

3. Is a maintenance manual available, or an operator's manual with a detailed description of recommended maintenance? Is the machine in good condition and well maintained? What changes/improvements in maintenance would improve it?

4. Separation results:

Spout	Type of material	Test weight	% Germination	% Purity	Weight	% of original lot

5. Draw a cutaway view of the spiral separator, showing how each fraction is separated.

6. Was the desired separation accomplished? Does the seed lot now meet standards? What further conditioning is required?

7. Was much good seed lost with the separated waste product? How could this loss be reduced or prevented? How can the lost good seed be recovered?

8. Make significant changes in the rate of seed feed as indicated, and note the results. (After noting one adjustment, return it to normal operation before changing the next adjustment.)

   a. Increase seed feed rate

   b. Decrease seed feed rate

9. What was done "to polish the spirals" of the spiral separator before the seed separation was begun? Why?

10. Points, conditions, or elements of special interest.

chapter fifty-eight

Width and thickness separator characteristics

Conditioner name:

Date:

Seed conditioning plant/machine location:

Machine brand, model number, other identification:

ear machine installed; condition:

Installation and feeding seed/waste materials to/from the machine description:

Any problems encountered with operation of the machine:

Procedure: Study the width and thickness separator and provide the following information or answers to questions.

1. Seed separating principle(s) used:

   a. Width

   b. Thickness

2. Describe how the separation is made.

3. On what crop seed is the width and thickness separator used? Why?

4. From where does the crop seed come as it goes to the width and thickness separator? Why?

5. Where does the crop seed go as it leaves the width and thickness separator? Why?

6. Is an operator's manual available? What is the machine designed to do? Is it being used for its designed purpose? If not, how is it being used? Can its use be changed to take better advantage of its design features? Is another machine(s) needed to improve cleaning performance?

Component, etc.	Needed for this machine	Comments
a. Elevator bringing seed not previously conditioned on this machine to the holding bin serving this machine. This elevator should have a two-way valve on its discharge, so that seed may bypass this machine if it is not needed.	Needed	To fit the machine's installation
b. Spouting of seed (coming from the previous separator) to this elevator.	Needed	To fit the machine's installation
c. Holding bin mounted over the feed intake of this machine.	Needed	To fit the machine's installation
d. Spouting of seed from this elevator to the holding bin mounted over this machine.	Needed	To fit the machine's installation
e. Spouting or flow of seed from the bin into the feed intake of this machine.	Needed	To fit the machine's installation
f. Mounting frame to support this machine in a position and at a height required for efficient operation.	Needed	To fit the machine's installation
g. Worker platforms, with required access steps or ladders, in appropriate locations so that operators can easily reach all controls of the machine for adjustments, and all parts of the machine for clean-out, setup, and repair.	Needed	To fit the machine's installation
h. Spouting or flow system to take good seed leaving this machine to the bin or conveyor/elevator taking good seed to the next machine in the conditioning sequence.	Needed	To fit the machine's installation
i. Dust and dusty air exhaust ducting and outside collector(s) handling dusty air and/or light materials separated from the good seed by this machine.	As needed	If needed, to fit the machine's installation
j. If the air around this machine is dusty, an exhaust fan/ducting/collection system should be installed to remove all dusty air.	As needed	If needed, to fit the machine's installation
k. Spouting to take each fraction of removed undesirable materials either to (1) bag collecting this material at this conditioning machine or to (2) a conveyor system taking all fractions of separated undesirable materials to a central waste product collecting system.	Needed	To fit the machine's installation
l. Elevator and required spouting to collect middling fraction(s) discharged from this machine back into the good-seed bin serving this machine, for reconditioning.	Not usually needed	Not needed

7. What are the usual uses of the width and thickness separator in conditioning seed? What other uses are possible?

8. What conditioning is usually done to a seed lot before it goes to the width and thickness separator? Why?

9. What is the usual position of the width and thickness separator in the conditioning line or sequence?

   a. Make and model.

   b. Name and address of manufacturer.

   c. Draw a simple diagram to illustrate the flow of seed into and through the separator and all separated fractions discharged from the separator. Label all seed fractions and essential parts of the separator.

10. What are the width and thickness separator's requirements for

    a. Power (including RPM)?

    b. Mounting and installation?

c. Seed supply and feed?

d. Handling discharged materials?

11. Overall dimensions:

Height

Width

Length

Height, floor to seed feed intake

Height, floor to lowest seed discharge

12. Draw the mounting base of the separator (i.e. the part that sits on the floor). Show all dimensions, and position of seed intake and all discharge spouts.

13. List and describe all adjustments, changes, etc., that can be made on the width and thickness separator, to affect its separation.

14. a. Describe the type of perforation used to make a **thickness** separation, and how the separation is made.

    b. Draw simple diagrams to illustrate the **thickness** separation perforation, and how the separation is made.

15. a. Describe the type of perforation used to make a **width** separation, and how the separation is made.

    b. Draw simple diagrams to illustrate the **width** separation perforation, and how the separation is made.

16. How is the seed feed rate controlled?

17. How is a uniform feed rate supplied to each first cylinder or screen section, if the separator has more than one unit performing the same separation?

18. Describe the mechanism (flat screens, cylindrical screens, etc.) providing the perforated separating surface, and how it functions.

19. What kind of motion is imparted to the perforated separating surface to enable it to make the separation? Describe fully both the motion and its mechanism.

20. What is the RPM of the separating cylinder or oscillating drive of the screen? Can RPM be changed or adjusted to affect the separation? How?

21. Describe the mechanism that keeps the perforations free of jammed seed, that is, keeps the perforations open. How does it work?

22. How many screens/cylinders are provided in this machine? How many separations does it make? How can additional separations be made?

23. Describe the procedures for changing screens/cylinders to make a different separation. How much time is required to make the change?

a. Where is seed handled by this machine coming from, and where is it being fed to?

b. What additional facilities, structures, subsidiary equipment, etc., are required for this machine to operate effectively and efficiently?

24. How are separated fractions discharged?

25. Describe the sequence of initial adjustments for setting up and operating the width and thickness separator.

26. Briefly discuss and describe other makes, models, and kinds of width and thickness separators available.

27. Points, conditions, or elements of special interest.

28. In the notebook that carries your study manual and your own observations, attach a copy of the operation and maintenance manuals of this model of machine that is installed in your facility.

29. Identify and describe five factors or conditions that influence the performance of this machine.

chapter fifty-nine

Width and thickness separator operation

Conditioner name:

Date:

Seed conditioning plant/machine location:

Machine brand, model number, other identification:

Year machine installed; condition:

Installation and feeding seed/waste materials to/from the machine description:

Any problems encountered with operation of the machine:

Procedure: Using a seed lot, set up, adjust, and operate the width and thickness separator as described below. Provide the following information or answers to questions.

1. Seed crop and variety.

   Seed lot:

   Weight

   Test weight

   % Germination

   % Purity

   Purpose, and kind, of separation desired

   Previous conditioning done on this seed lot

2. Make, model, and manufacturer of the width and thickness separator.

3. Is a maintenance manual available, or an operator's manual with a detailed description of recommended maintenance? Is the machine in good condition and well maintained? What changes/improvements in maintenance would improve it?

Component, etc.	Needed for this machine	Comments
a. Elevator bringing seed not previously conditioned on this machine to the holding bin serving this machine. This elevator should have a two-way valve on its discharge, so that seed may bypass this machine if it is not needed.	Needed	To fit the machine's installation
b. Spouting of seed (coming from the previous separator) to this elevator.	Needed	To fit the machine's installation
c. Holding bin mounted over the feed intake of this machine.	Needed	To fit the machine's installation
d. Spouting of seed from this elevator to the holding bin mounted over this machine.	Needed	To fit the machine's installation
e. Spouting or flow of seed from the bin into the feed intake of this machine.	Needed	To fit the machine's installation
f. Mounting frame to support this machine in a position and at a height required for efficient operation.	Needed	To fit the machine's installation
g. Worker platforms, with required access steps or ladders, in appropriate locations so that operators can easily reach all controls of the machine for adjustments, and all parts of the machine for clean-out, setup, and repair.	Needed	To fit the machine's installation

Continued

Component, etc.	Needed for this machine	Comments
h. Spouting or flow system to take good seed leaving this machine to the bin or conveyor/elevator taking good seed to the next machine in the conditioning sequence.	Needed	To fit the machine's installation
i. Dust and dusty air exhaust ducting and outside collector(s) handling dusty air and/or light materials separated from the good seed by this machine.	As needed	If needed, to fit the machine's installation
j. If the air around this machine is dusty, an exhaust fan/ducting/collection system should be installed to remove all dusty air.	As needed	If needed, to fit the machine's installation
k. Spouting to take each fraction of removed undesirable materials either to (1) bag collecting this material at this conditioning machine or to (2) a conveyor system taking all fractions of separated undesirable materials to a central waste product collecting system.	Needed	To fit the machine's installation
l. Elevator and required spouting to collect middling fraction(s) discharged from this machine back into the good-seed bin serving this machine, for reconditioning.	Not usually needed	Not needed

4. Approximate adjustments used.

Adjustment	Setting used

5. Separation results.

Spout	Type of material	Test weight	% Germination	% Purity	Weight	% of original lot

6. Size evaluation

	Thickness				Width			
	Minimum size		Maximum size		Minimum size		Maximum size	
Seed fraction	%	Goes through screen	%	Remains on top of screen	%	Goes through screen	%	Remains on top of screen
Original lot								
Separated fraction								
1								
2								
3								
4								
5								
6								

7. Was the desired separation accomplished? Does the seed lot now meet standards? What further conditioning is required?

8. Make significant changes in adjustments as indicated below, and note the results. (After noting one adjustment, return it to normal operation before changing the next adjustment.)

   a. Increase feed rate.

   b. Decrease feed rate.

   c. Change the position (describe) of separating screens or cylinder shells.

9. Diagram the flow of seed through the separator. Label each fraction.

10. Points, conditions, or elements of special interest.

chapter sixty

Roll mill characteristics

Conditioner name:

Date:

Seed conditioning plant/machine location:

Machine brand, model number, other identification:

Year machine installed; condition:

Installation and feeding seed/waste materials to/from the machine description:

Any problems encountered with operation of the machine:

Procedure: Study the roll mill and provide the following information or answers to questions.

1. The roll mill separates seed by what physical differences in seed?

   Describe its separating action.

2. How, and for what purposes, is the roll mill used in seed conditioning?

3. On what crop seed is the roll mill used? Why?

4. From where does the crop seed come as it goes to the roll mill? Why?

5. Where does the crop seed go as it leaves the roll mill?

6. Is an operator's manual available? What is the machine designed to do? Is it being used for its designed purpose? If not, how is it being used? Can its use be changed to take better advantage of its design features? Is another machine(s) needed to improve cleaning performance?

7. What conditioning is done on a seed lot before it goes to the roll mill? Why?

8. At what point in the seed conditioning line or sequence is the roll mill installed and used? Why is it installed in this position?

9. Make and model.

10. Name and address of manufacturer.

11. Overall dimensions:

    Height

    Width

    Length

Height, floor to seed feed intake

Height, floor to good seed discharge

12. What are the requirements of the roll mill for

    a. Power (including RPM)?

    b. Seed supply and feed?

    c. Mounting and installation?

    d. Handling discharged good seed?

    e. Handling discharged waste and middling products?

13. Draw the roll mill's mounting base (i.e., the part which sits on the floor). Show all dimensions, position of the seed feed intake, good seed discharge spout, and all waste product spouts.

14. How many pairs of rolls does this model have?

15. List and describe all adjustments on the roll mill.

16. Why is a pair (not one, or single) of rolls required to make a seed separation?

17. Illustrate a pair of rolls.

    Side (or perspective) view

    End view

18. Does seed flow from one pair of rolls onto the next, etc., or does each pair of rolls operate separately and independently? Explain.

19. Illustrate the position of all pairs of rolls, and show the flow of seed into, through, and out of the roll mill.

20. What is the purpose of having more than one pair of rolls?

21. What is the material covering the rolls? What is the resulting surface texture of the rolls? Why is the surface texture important?

22. How is each pair of rolls driven? How many motors does the roll mill have?

23. Can the speed of rotation of the rolls be changed? How? Why? Is the speed of each pair controlled separately, or do all revolve at the same speed?

24. How is the rate of seed feed adjusted? How are all pairs of rolls fed uniformly?

25. Where and how is the seed mixture (to be separated) fed to the rolls? Describe.

26. What is the purpose of the small gate at the bottom of the feed column?

27. Illustrate the position of the shield above a pair of rolls.

28. a. What is the function of the shield (above each pair of rolls) in making a separation?

    b. Illustrate the shields and their function.

29. Can the position of the shields be adjusted? How? Why?

30. a. Can the angle of inclination of the rolls be adjusted? How? Why?

    b. Illustrate with a simple drawing.

31. a. How many discharge spouts does the roll mill have?

    b. With a simple drawing, illustrate the position of all discharge spouts.

    c. What discharges from each spout?

32. Describe the sequence of initial adjustments for setting up and operating the roll mill.

33. Briefly discuss and describe other makes, models, and kinds of roll mill available.

Component, etc.	Needed for this machine	Comments
a. Elevator bringing seed not previously conditioned on this machine to the holding bin serving this machine. This elevator should have a two-way valve on its discharge, so that seed may bypass this machine if it is not needed.	Needed	To fit the machine's installation
b. Spouting of seed (coming from the previous separator) to this elevator.	Needed	To fit the machine's installation
c. Holding bin mounted over the feed intake of this machine.	Needed	To fit the machine's installation
d. Spouting of seed from this elevator to the holding bin mounted over this machine.	Needed	To fit the machine's installation
e. Spouting or flow of seed from the bin into the feed intake of this machine.	Needed	To fit the machine's installation
f. Mounting frame to support this machine in a position and at a height required for efficient operation.	Needed	To fit the machine's installation
g. Worker platforms, with required access steps or ladders, in appropriate locations so that operators can easily reach all controls of the machine for adjustments, and all parts of the machine for clean-out, setup, and repair.	Needed	To fit the machine's installation
h. Spouting or flow system to take good seed leaving this machine to the bin or conveyor/elevator taking good seed to the next machine in the conditioning sequence.	Needed	To fit the machine's installation
i. Dust and dusty air exhaust ducting and outside collector(s) handling dusty air and/or light materials separated from the good seed by this machine.	As needed	If needed, to fit the machine's installation
j. If the air around this machine is dusty, an exhaust fan/ducting/collection system should be installed to remove all dusty air.	As needed	If needed, to fit the machine's installation
k. Spouting to take each fraction of removed undesirable materials either to (1) bag collecting this material at this conditioning machine or to (2) a conveyor system taking all fractions of separated undesirable materials to a central waste product collecting system.	Needed	To fit the machine's installation
l. Elevator and required spouting to collect middling fraction(s) discharged from this machine back into the good-seed bin serving this machine, for reconditioning.	Not usually needed	Not needed

a. Where is seed handled by this machine coming from, and where is it being fed to?

b. What additional facilities, structures, subsidiary equipment, etc., are required for this machine to operate effectively and efficiently?

34. Points, conditions, or elements of special interest.

35. In the notebook that carries your study manual and your own observations, attach a copy of the operation and maintenance manuals of this model of machine that is installed in your facility.

36. Identify and describe five factors or conditions which influence the performance of this machine.

chapter sixty-one

Roll mill operation

Conditioner name:

Date:

Seed conditioning plant/machine location:

Machine brand, model number, other identification:

Year machine installed; condition:

Installation and feeding seed/waste materials to/from the machine description:

Any problems encountered with operation of the machine:

Procedure: Using a seed lot, set up, adjust, and operate the roll mill as described. Provide the following information or answers to questions.

1. Crop seed and variety.

   Seed lot:

   Weight

   Test weight

   % Purity

   % Germination

   Impurities

2. Purpose of using roll mill.

3. Is a maintenance manual available, or an operator's manual with a detailed description of recommended maintenance? Is the machine in good condition and well maintained? What changes/improvements in maintenance would improve it?

4. Conditioning already done on this seed lot.

5. Roll mill make, model, and manufacturer.

6. Approximate adjustments used.

Adjustment	Setting used

7. a. Separation results.

Spout	Type of material	Test weight	% Germination	% Purity	Weight	% of original lot

b. Illustrate the position of the discharge spouts and indicate the material that discharges from each.

8. How can good seed in the mixtures discharging from spouts no. 2 and no. 3 be recovered?

9. Was the desired separation accomplished? Does the seed lot now meet standards? What further conditioning is required?

10. Make significant changes, as indicated, in each of the following adjustments, and note the results. (Return the adjustment to normal operation before changing the next adjustment.)

    a. Increase seed feed rate

    b. Decrease seed feed rate

    c. Increase roll speed

    d. Decrease roll speed

    e. Increase slope of rolls

    f. Decrease slope of rolls

    g. Raise shields

    h. Lower shields

11. Points, conditions, or elements of special interest.

chapter sixty-two

Color separator characteristics

Conditioner name:

Date:

Seed conditioning plant/machine location:

Machine brand, model number, other identification:

Year machine installed; condition:

Installation and feeding seed/waste materials to/from the machine description:

Any problems encountered with operation of the machine:

Procedure: Study the color separator and provide the following information or answers to questions.

1. Seed separating principle(s) used.

2. What are the usual uses of the color separator in seed conditioning? What other uses are possible?

3. On what crop seed is the color separator used? Why?

4. From where does the crop seed come as it goes to the color separator? Why?

5. Where does the crop seed go as it leaves the color separator?

6. Is an operator's manual available? What is the machine designed to do? Is it being used for its designed purpose? If not, how is it being used? Can its use be changed to take better advantage of its design features? Is another machine(s) needed to improve cleaning performance?

7. What is the usual position of the color separator in the conditioning line or sequence?

8. Where is seed handled by this machine coming from, and where is it being fed to?

9. What additional facilities, structures, subsidiary equipment, etc., are required for this machine to operate effectively and efficiently?

10. What conditioning is usually done to a seed lot before it goes to the color separator? Why?

11. Describe how a separation is made by the color separator.

12. Make and model.

13. Name and address of manufacturer.

14. Draw a simple diagram to illustrate seed flow into and through the color separator as the separation is made, and how all separated fractions are discharged. Label all seed fractions and essential parts of the separator.

15. What are the color separator's requirements for

   a. Power (describe fully)?

   b. Installation and mounting?

   c. Seed supply and feed?

   d. Handling discharged fractions?

   e. Dust control?

   f. Operator competence and training?

   g. Compressed air?

16. Overall dimensions:

    Height

    Width

    Length

    Height, floor to seed intake

    Height, floor to good seed discharge

17. Draw the color separator's mounting base, showing all dimensions. Indicate the position of the seed feed intake and all discharge spouts.

18. List and describe all adjustments, changes, etc., that can be made to affect the separation.

19. Describe the circuitry system. How can circuits be changed or repaired?

20. List and briefly describe the function and operation of each major system of circuits.

21. When a circuit malfunctions, how is this known?

22. How is seed fed into the separator? How is feed rate adjusted?

23. Is dust removed from the seed lot before it enters the separation chamber? Describe.

24. How many separation chambers does the separator have? Can each function separately? Describe.

25. Describe the photocell system in the inspection chamber, and how it functions.

26. Describe the ejector system, and how it removes undesirable seed.

27. How many ejection air pulses can be delivered per second for ejection of undesirable seed? How does this affect capacity?

28. Describe the reference, background, or adjustment color system, how it is used with the separator, and how the correct color is selected.

29. Does the separator have its own air compressor? Describe the system, air pressure, and air delivery.

30. How is dust and trash kept out of the compressed air system?

31. What provisions are made for clean-out of the separator? How often must it be cleaned out? How is it cleaned out?

32. Can the separator be operated under ambient conditions, or is a special environment required? Describe.

33. Can several separators be used in a parallel-flow "bank" to increase capacity? Describe.

34. Describe the sequence of initial adjustments for setting up and operating the color separator.

35. Briefly describe other makes and models of color separators available.

36. Points, conditions, or elements of special interest.

37. In the notebook that carries your study manual and your own observations, attach a copy of the operation and maintenance manuals of this model of machine that is installed in your facility.

38. Identify and describe five factors or conditions that influence the performance of this machine.

chapter sixty-three

Color separator operation

Conditioner name:

Date:

Seed conditioning plant/machine location:

Machine brand, model number, other identification:

Year machine installed; condition:

Installation and feeding seed/waste materials to/from the machine description:

Any problems encountered with operation of the machine:

Procedure: Using a seed lot, set up, adjust, and operate the color separator as described. Provide the following information or answers to questions.

1. Crop seed and variety.

   Seed lot:

   Weight

   Test weight

   % Purity

   % Germination

   Impurities

2. Purpose of using color separator

3. Conditioning already done on this seed lot

4. Is a maintenance manual available, or an operator's manual with a detailed description of recommended maintenance? Is the machine in good condition and well maintained? What changes/improvements in maintenance would improve it?

5. Make, model, and manufacturer of color separator.

6. Approximate adjustments used.

Adjustment	Setting used

7. Describe the procedure used to determine whether adjustments were correct.

8. Separation results.

Spout	Type of material	Test weight	% Germination	% Purity	Weight	% of original lot

9. Was the desired separation accomplished? Does the seed lot now meet standards? What further conditioning is required?

10. Was much good seed lost with the undesirable materials separated? How could this loss be reduced? How could the lost good seed be recovered?

11. Make significant changes in adjustments as indicated below, and note the results. (After noting one adjustment, return it to normal operation before changing the next adjustment.)

    a. Increase feed rate

    b. Decrease feed rate

    c. Use darker reference background

    d. Use lighter reference background

    e. Increase light sensitivity

    f. Decrease light sensitivity

    g. Increase dark sensitivity

    h. Decrease dark sensitivity

    i. Other adjustments (describe)

12. Was much dust created? Does dust affect the separation? How can dust be reduced and/or controlled?

13. Using seed samples, determine the reference color backgrounds and approximate separator settings to make the desired separations on several samples.

Sample no.	Crop and variety	Contaminant(s) to be removed	Reference color background	Setting
1				
2				
3				
4				
5				

14. Points, conditions, or elements of special interest.

chapter sixty-four

Magnetic separator characteristics

Conditioner name:

Date:

Seed conditioning plant/machine location:

Machine brand, model number, other identification:

Year machine installed; condition:

Installation and feeding seed/waste materials to/from the machine description:

Any problems encountered with operation of the machine:

Procedure: Study the magnetic separator and provide the following information or answers to questions.

1. The magnetic separator separates seed by what physical differences of the seed?

2. Explain how the separation is made.

3. What conditioning is done to a seed lot before it goes to the magnetic separator? Why?

4. On what crop seed is the magnetic separator used? Why?

5. From where does the crop seed come as it goes to the magnetic separator? Why?

6. Where does the crop seed go as it leaves the magnetic separator? Why?

7. Is an operator's manual available? What is the machine designed to do? Is it being used for its designed purpose? If not, how is it being used? Can its use be changed to take better advantage of its design features? Is another machine(s) needed to improve cleaning performance?

8. How is the magnetic separator used in seed conditioning?

Component, etc.	Needed for this machine	Comments
a. Elevator bringing seed not previously conditioned on this machine to the holding bin serving this machine. This elevator should have a two-way valve on its discharge, so that seed may bypass this machine if it is not needed.	Needed	To fit the machine's installation
b. Spouting of seed (coming from the previous separator) to this elevator.	Needed	To fit the machine's installation
c. Holding bin mounted over the feed intake of this machine.	Needed	To fit the machine's installation
d. Spouting of seed from this elevator to the holding bin mounted over this machine.	Needed	To fit the machine's installation
e. Spouting or flow of seed from the bin into the feed intake of this machine.	Needed	To fit the machine's installation
f. Mounting frame to support this machine in a position and at a height required for efficient operation.	Needed	To fit the machine's installation
g. Worker platforms, with required access steps or ladders, in appropriate locations so that operators can easily reach all controls of the machine for adjustments, and all parts of the machine for clean-out, setup, and repair.	Needed	To fit the machine's installation
h. Spouting or flow system to take good seed leaving this machine to the bin or conveyor/elevator taking good seed to the next machine in the conditioning sequence.	Needed	To fit the machine's installation
i. Dust and dusty air exhaust ducting and outside collector(s) handling dusty air and/or light materials separated from the good seed by this machine.	As needed	If needed, to fit the machine's installation
j. If the air around this machine is dusty, an exhaust fan/ducting/collection system should be installed to remove all dusty air.	As needed	If needed, to fit the machine's installation
k. Spouting to take each fraction of removed undesirable materials either to (1) bag collecting this material at this conditioning machine or to (2) a conveyor system taking all fractions of separated undesirable materials to a central waste product collecting system.	Needed	To fit the machine's installation
l. Elevator and required spouting to collect middling fraction(s) discharged from this machine back into the good-seed bin serving this machine, for reconditioning.	Not usually needed	Not needed

9. At what point in the seed conditioning sequence or line is the magnetic separator installed and used? Why is it installed at this position?

10. Make and model of magnetic separator.

11. Name and address of manufacturer.

12. What are the requirements of the magnetic separator for

    a. Installation and mounting?

    b. Seed supply and feed?

    c. Handling separated fractions?

    d. Electric power?

13. Overall dimensions:

    Height

    Width

    Length

    Height, floor to seed feed intake

    Height, floor to good seed discharge

14. Where is seed handled by this machine coming from, and where is it being fed to?

15. What additional facilities, structures, subsidiary equipment, etc., are required for this machine to operate effectively and efficiently?

16. Draw the separator's mounting base (i.e., the part that sits on the floor). Show all dimensions, position of the seed feed or intake, good seed discharge spout, and waste product discharge spouts.

17. Illustrate seed flow into, through, and out of the magnetic separator.

18. List and describe the adjustments or changes that can be made to control the separation made by the magnetic separator.

19. What is added to the seed mixture before it is separated by the magnetic separator? Why is each added component necessary?

20. How are the components described in 13 above added to the seed mixture?

    a. Describe fully, including sequence, amount, etc.

    b. Illustrate.

21. Can one or more of the added components described above be omitted? Which? Explain.

22. Is the mixing process (see questions above) continuous or in batches? Describe.

23. a. From what is the magnetic powder made?

    b. Describe the physical characteristics (texture, fineness, etc.) of the magnetic powder.

    c. Describe the electrical and chemical characteristics of the magnetic powder.

24. a. To what kinds of seed surfaces will the magnetic powder adhere? Describe.

    b. What must be done to the surface of seed before the magnetic powder will adhere to the seed? Explain.

25. To what kinds of seed surfaces will the magnetic powder NOT adhere? Describe.

26. Can excess magnetic powder used in the separation be recovered and reused (i.e., used more than once)? How? Why?

27. a. How many magnetic drums does this separator have?

    b. Drum width, diameter, and RPM.

28. Seed flow

    a. Is each drum fed separately, or does the same seed flow consecutively over all drums? Describe seed flow.

    b. Illustrate the seed flow with a simple drawing.

29. Can speed of the drums be adjusted? How? Why?

30. Can the seed feed rate be adjusted? How?

31. Describe how all separated seed fractions discharge from and/or are removed from the magnetic drum.

32. What is the source of magnetism?

33. If electromagnets are used, how is electric energy provided to the magnets? Can the magnetism be varied? Describe.

34. How is the magnetic separator cleaned out?

35. Describe the initial sequence of adjustments for setting up and operating the magnetic separator.

36. Briefly describe and discuss other makes, models, and kinds of magnetic separators available.

37. Points, conditions, or elements of special interest.

38. In the notebook that carries your study manual and your own observations, attach a copy of the operation and maintenance manuals of this model of machine that is installed in your facility.

39. Identify and describe five factors or conditions that influence the performance of this machine.

chapter sixty-five

Magnetic separator operation

Conditioner name:

Date:

Seed conditioning plant/machine location:

Machine brand, model number, other identification:

Year machine installed; condition:

Installation and feeding seed/waste materials to/from the machine description:

Any problems encountered with operation of the machine:

Procedure: Using a suitable seed lot, set up, adjust, and operate the magnetic separator as described. Provide the following information or answers to questions.

1. Crop seed and variety.

   Seed lot:

   - Weight
   - Test weight
   - % Purity
   - % Germination
   - Impurities

   Impurities to be removed

   Conditioning already done on this seed lot

2. Make, model, and manufacturer of magnetic separator.

3. Is a maintenance manual available, or an operator's manual with a detailed description of recommended maintenance? Is the machine in good condition and well maintained? What changes/improvements in maintenance would improve it?

4. Where is seed handled by this machine coming from, and where is it being fed to?

5. What additional facilities, structures, subsidiary equipment, etc., are required for this machine to operate effectively and efficiently?

6. a. Separating components added.

Component	Description	Amount added	% of seed weight	Remarks

b. Sequence of adding separation components.

7. a. Describe the process of adding and mixing the separation components.

b. Describe the appearance of seed to which the magnetic powder did *not* adhere.

8. Was the desired separation accomplished? Does the seed lot now meet standards? What further conditioning is required?

9. Make significant changes in adjustments as indicated below, and note the results. (After noting one adjustment, return it to normal operation before changing the next adjustment.)

   a. Increase feed rate

   b. Decrease feed rate

   c. Increase magnetic force (on electromagnet only)

   d. Decrease magnetic force (on electromagnet only)

   e. Increase speed of magnet drum

   f. Decrease speed of magnet drum

10. Points, conditions, or elements of special interest.

chapter sixty-six

Electrostatic separator characteristics

Conditioner name:

Date:

Seed conditioning plant/machine location:

Machine brand, model number, other identification:

Year machine installed; condition:

Installation and feeding seed/waste materials to/from the machine description:

Any problems encountered with operation of the machine:

Procedure: Study the electrostatic separator and provide the following information or answers to questions.

1. Seed separating principle(s) used.

2. Describe the uses and limitations of the electrostatic separator in seed conditioning.

3. On what crop seed is the electrostatic separator used? Why?

4. From where does the crop seed come as it goes to the electrostatic separator? Why?

5. Where does the crop seed go as it leaves the electrostatic separator?

6. Is an operator's manual available? What is the machine designed to do? Is it being used for its designed purpose? If not, how is it being used? Can its use be changed to take better advantage of its design features? Is another machine(s) needed to improve cleaning performance?

7. What is the usual location of the electrostatic separator in the conditioning sequence? What conditioning is done to a seed lot before it goes to the electrostatic separator? Why?

Component, etc.	Needed for this machine	Comments
a. Elevator bringing seed not previously conditioned on this machine to the holding bin serving this machine. This elevator should have a two-way valve on its discharge, so that seed may bypass this machine if it is not needed.	Needed	To fit the machine's installation
b. Spouting of seed (coming from the previous separator) to this elevator.	Needed	To fit the machine's installation
c. Holding bin mounted over the feed intake of this machine.	Needed	To fit the machine' installation
d. Spouting of seed from this elevator to the holding bin mounted over this machine.	Needed	To fit the machine' installation
e. Spouting or flow of seed from the bin into the feed intake of this machine.	Needed	To fit the machine's installation
f. Mounting frame to support this machine in a position and at a height required for efficient operation.	Needed	To fit the machine's installation
g. Worker platforms, with required access steps or ladders, in appropriate locations so that operators can easily reach all controls of the machine for adjustments, and all parts of the machine for clean-out, setup, and repair.	Needed	To fit the machine's installation
h. Spouting or flow system to take good seed leaving this machine to the bin or conveyor/elevator taking good seed to the next machine in the conditioning sequence.	Needed	To fit the machine's installation
i. Dust and dusty air exhaust ducting and outside collector(s) handling dusty air and/or light materials separated from the good seed by this machine.	As needed	If needed, to fit the machine's installation
j. If the air around this machine is dusty, an exhaust fan/ducting/collection system should be installed to remove all dusty air.	As needed	If needed, to fit the machine's installation
k. Spouting to take each fraction of removed undesirable materials either to (1) bag collecting this material at this conditioning machine or to (2) a conveyor system taking all fractions of separated undesirable materials to a central waste product collecting system.	Needed	To fit the machine's installation
l. Elevator and required spouting to collect middling fraction(s) discharged from this machine back into the good-seed bin serving this machine, for reconditioning.	Not usually needed	Not needed

8. Describe the changes in electrical charges on (or in) the seed that permit an electrostatic separation.

9. a. What are the two different types of separation made by the electrostatic separator?

   b. Describe each of the two types of separation.

10. Make and model.

11. Name and address of manufacturer.

12. Draw a simple diagram to illustrate seed flow into and through the electrostatic separator, and how all separated fractions are discharged. Label all seed fractions and essential parts of the separator.

13. What are the electrostatic separator's requirements for

    a. Power?

    b. Seed feed, supply, and preseparation preparation?

c. Installation and mounting?

d. Handling discharged clean seed and waste materials?

14. Overall dimensions:

Height

Length

Width

Height to seed intake

Height to good seed discharge

15. Where is seed handled by this machine coming from, and where is it being fed to?

16. What additional facilities, structures, subsidiary equipment, etc., are required for this machine to operate effectively and efficiently?

17. Draw the mounting base (i.e., the part that sits on the floor) of the electrostatic separator. Show all dimensions, and the position of seed feed intake and all discharge spouts.

18. a. Draw the discharge spouts of the electrostatic separator.

    b. Can the position of the discharge spouts be adjusted (i.e. the seed that goes into the spouts)? Which ones? How? Why?

19. List and describe all adjustments, changes, etc., that can be made to affect the separation.

20. How are seed conveyed into the electrostatic field?

21. How is rate of seed feed adjusted?

22. Can polarity (positive or negative) of the electrostatic field be changed? How? Why?

23. The voltage going to the electrode can be varied from ________ to ______ volts (AC)/(DC). How is this voltage supplied?

24. a. What is the width of the mechanism (roller, etc.) by which the seed stream is conveyed into the electrostatic field?

    b. What is the length of the electrode?

25. Describe the electrode.

26. Can the position of the electrode be adjusted? How? Why?

27. How are seed that stick to the conveying roller or belt removed?

28. Sometimes only a few of the weed seed (or contaminant to be removed) can be taken out in one pass through the separator. How can a complete separation be made?

29. Describe the sequence of initial adjustments for setting up and operating the electrostatic separator.

30. Briefly describe and discuss other makes, models, and kinds of electrostatic separators available.

31. Points, conditions, or elements of special interest.

32. In the notebook that carries your study manual and your own observations, attach a copy of the operation and maintenance manuals of this model of machine that is installed in your facility.

33. Identify and describe five factors or conditions that influence the performance of this machine.

chapter sixty-seven

Electrostatic separator operation

Conditioner name:

Date:

Seed conditioning plant/machine location:

Machine brand, model number, other identification:

Year machine installed; condition:

Installation and feeding seed/waste materials to/from the machine description:

Any problems encountered with operation of the machine:

Procedure: Using a seed lot, set up, adjust, and operate the electrostatic separator as described. Provide the following information or answers to questions

1. Seed lot:

   Weight

   Test weight

   % Purity

   % Germination

   Impurities

2. Purpose of using color separator.

3. Conditioning already done on this seed lot.

4. Make, model, and manufacturer of color separator.

5. Is a maintenance manual available, or an operator's manual with a detailed description of recommended maintenance? Is the machine in good condition and well maintained? What changes/improvements in maintenance would improve it?

6. Approximate adjustments used.

Adjustment	Setting used

7. Separation results.

Spout	Type of material	Test weight	% Germination	% Purity	Weight	% of original lot

8. Was the desired separation accomplished? Does the seed lot now meet standards? What further conditioning is required? Could the seed lot be improved by additional passes through the electrostatic separator?

9. Was much good seed lost with the undesirable materials separated? How could this loss be reduced? How could the lost good seed be recovered?

10. Make significant changes in adjustments as indicated below, and note the results. (After noting one adjustment, return it to normal operation before changing the next adjustment.)

    a. Increase feed rate

    b. Decrease feed rate

c. Increase speed of roller or belt

d. Decrease speed of roller or belt

e. Move electrode closer to seed

f. Move electrode farther from seed

g. Increase voltage

h. Decrease voltage

i. Change polarity of charge (specify)

j. Alter seed condition (heat, dry, add moisture, etc.; specify)

11. Points, conditions, or elements of special interest.

chapter sixty-eight

Seed treater characteristics

Conditioner name:

Date:

Seed conditioning plant/machine location:

Machine brand, model number, other identification:

Year machine installed; condition:

Installation and feeding seed/waste materials to/from the machine description:

Any problems encountered with operation of the machine:

Procedure: Study the seed treater and treatment materials. Provide the following information or answers to questions.

1. Why are seed treated?

2. Should seed of a disease-resistant variety be treated? Explain.

3. Seed may be treated with

   a. A FUNGICIDE to

      (1)

      (2)

      (3)

      (4)

   b. An INSECTICIDE to

      (1)

      (2)

      (3)

      (4)

Component, etc.	Needed for this machine	Comments
a. Elevator bringing seed not previously conditioned on this machine to the holding bin serving this machine. This elevator should have a two-way valve on its discharge, so that seed may bypass this machine if it is not needed.	Needed	To fit the machine's installation
b. Spouting of seed (coming from the previous separator) to this elevator.	Needed	To fit the machine's installation
c. Holding bin mounted over the feed intake of this machine.	Needed	To fit the machine's installation
d. Spouting of seed from this elevator to the holding bin mounted over this machine.	Needed	To fit the machine's installation
e. Spouting or flow of seed from the bin into the feed intake of this machine.	Needed	To fit the machine's installation
f. Mounting frame to support this machine in a position and at a height required for efficient operation.	Needed	To fit the machine's installation
g. Worker platforms, with required access steps or ladders, in appropriate locations so that operators can easily reach all controls of the machine for adjustments, and all parts of the machine for clean-out, setup, and repair.	Needed	To fit the machine's installation
h. Spouting or flow system to take good seed leaving this machine to the bin or conveyor/elevator taking good seed to the next machine in the conditioning sequence.	Needed	To fit the machine's installation
i. Dust and dusty air exhaust ducting and outside collector(s) handling dusty air and/or light materials separated from the good seed by this machine.	As needed	If needed, to fit the machine's installation
j. If the air around this machine is dusty, an exhaust fan/ducting/collection system should be installed to remove all dusty air.	As needed	If needed, to fit the machine's installation
k. Spouting to take each fraction of removed undesirable materials either to (1) bag collecting this material at this conditioning machine or to (2) a conveyor system taking all fractions of separated undesirable materials to a central waste product collecting system.	Needed	To fit the machine's installation
l. Elevator and required spouting to collect middling fraction(s) discharged from this machine back into the good-seed bin serving this machine, for reconditioning.	Not usually needed	Not needed

4. The characteristics of a good seed treatment are

   a.

   b.

   c.

   d.

   e.

   f.

   g.

   h.

5. Seed treatments may be applied in either of three physical forms:

   a.

   b.

   c.

6. What is the difference between a slurry treatment and a liquid treatment?

7. Why do most seed treatment materials contain dyes?

8. Is it necessary to dry seed after treating with a slurry or liquid? Explain.

9. Describe safety precautions to avoid damaging seed in using seed treatments.

10. Describe safety precautions for handling treated seed to prevent accidental poisoning of workers or farm animals.

11. a. What are the advantages and disadvantages, from the point of view of the farmer and the commercial seedsman, of treating seed?

    b. Why are seed treated as part of the seed conditioning operation?

12. What is the position of the seed treater in the conditioning line or flow sequence? Why?

13. Treater make and model, and name and address of manufacturer.

14. Is an operator's manual available? What is the machine designed to do? Is it being used for its designed purpose? If not, how is it being used? Can its use be changed to take better advantage of its design features? Is another machine(s) needed to improve cleaning performance?

15. Overall dimensions:

    Height

    Length

    Width

    Height, floor to seed feed intake

    Height, floor to treated seed discharge

16. Where is seed handled by this machine coming from, and where is it being fed to?

17. What additional facilities, structures, subsidiary equipment, etc., are required for this machine to operate effectively and efficiently?

18. Draw the mounting base of the treater (i.e., the part that sits on the floor). Show all dimensions, and the position of seed intake, treated seed discharge, and air exhaust outlets.

19. What are the treater's requirements for

    a. Power?

    b. Seed supply and feed?

    c. Installation and mounting?

    d. Handling discharged seed?

    e. Auxiliary or separate component tanks, pumps, premixing, mixers, etc.?

    f. Water supply to the mixing tank?

    g. Dust control and air ducting?

20. This treater applies chemicals in what form(s)?

21. This treater can simultaneously apply ____________ separate materials to the seed.

22. Describe the treating operation of this treater.

23. Parts that contact the chemical are constructed of what material? Is it corrosion resistant?

24. Diagram the flow of seed and treatment material through the treater. Label all essential treater parts, and the flows of seed and chemical.

25. List and describe all adjustments that affect the capacity, rate of treatment dosage, and seed coverage of the treater.

26. How is seed feed controlled or adjusted? Can seed feed be stopped with the treater's feed control? Describe.

27. How can the amount of treatment applied to a given weight of seed be changed? Either of two distinct methods, or a combination of both, may be used; describe both.

28. How is the weight of seed measured, to ensure applying exact dosages?

29. How can the weight of seed metered into the treating chamber be changed or adjusted? Describe.

30. Describe the chemical metering system of the treater.

31. How is the treating chemical measured, to ensure exact treatment rates? Describe.

32. What sizes of metering cups (dippers) are available? How are they identified? How much liquid or solution slurry does each actually hold?

33. How does treating chemical flow from the metering tank to the treating chamber? Describe.

34. How is chemical added to the seed? Describe.

35. a. How are seed and chemical blended together to spread chemical over all seed?

b. Describe the blending system, blending chamber, and all parts.

c. Draw the blending chamber, illustrating how it functions. Label all essential parts and show all dimensions.

36. Can the flow of seed through the blending chamber be retarded or speeded up, to change the time that seed is subjected to the blending action? How?

37. How can seed discharging from the treater be handled? What is the major factor(s) influencing how treated seed will be handled?

38. What is the capacity of the main treatment chemical tank?

39. Does the tank have an agitator or other device to keep the chemical in solution (prevent settling out of the chemical powder)? Describe.

40. How far from the treater can the main treatment chemical supply tank be installed? Can it be installed below, above, or at the same level? Describe.

41. What kind of pump is used to move the treatment material? Where is it mounted? Where can or should it be mounted?

42. Does the treating solution circulate through the metering tank? Describe the circulating flow. Does the solution foam? How can this be prevented?

43. When the treater or pump is stopped, does all chemical drain out of the metering tank and return to the main mixing tank, or does some remain in the metering tank? Describe the system and construction used and its effect on solution remaining in the metering tank.

44. Describe the sequence of adjustments for setting up and operating the treater.

45. Briefly describe and discuss other makes, models, and kinds of seed treaters available.

46. Points, conditions, or elements of special interest.

47. In the notebook that carries your study manual and your own observations, attach a copy of the operation and maintenance manuals of this model of machine that is installed in your facility.

48. Identify and describe five factors or conditions that influence the performance of this machine.

chapter sixty-nine

Seed treater operation

Conditioner name:

Date:

Seed conditioning plant/machine location:

Machine brand, model number, other identification:

Year machine installed; condition:

Installation and feeding seed/waste materials to/from the machine description:

Any problems encountered with operation of the machine:

Procedure: Using a seed lot, calculate treating rate and set up, adjust, and operate the seed treater as described. Provide the following information or answers to questions.

THE SEED:

1. Seed crop and variety

   Seed lot weight

   Reasons for treating the seed

THE TREATMENT:

2. Reasons for treating the seed.
3. Treating material(s) used, and form in which the material comes from the supplier.
4. Treating rate(s) used.
5. Form in which the treatment(s) will be applied to the seed.
6. Treating rate(s) recommended.

7. a. Was the treating material diluted? With what? How much? Describe.

   b. Were a spreader, sticker, foam suppressor, etc., used? Why? Describe.

THE TREATER:

8. Treater make, model, and manufacturer.

9. Seed dump weight setting and weight of seed delivered per dump.

10. Chemical cup size(s) used, actual amount of material delivered by one cup, and number of cups that deliver chemical per one seed dump.

11. Describe the treater's mixing chamber and its operation.

12. a. How are seed supplied to the treater? Does it receive a uniform, even flow of seed?

    b. How are treated seed handled as they leave the treater? Does this interfere with treating operations or capacity, or is it adequate for the operating capacity?

13. How is the operating supply of treating material maintained and supplied to the treater (main or reserve tank and pump)?

14. How is the treating material (powder) kept properly in solution, if it has been diluted or is in slurry form?

CALCULATING TREATER SETTINGS:

15. Treatment rate desired ______________ cc's per ________________. (If treatment rate is expressed in units other than cubic centimeters (cc), convert before entering here, since most treater chemical cups measure sizes in cc).

16. Size or actual capacity of chemical treater cups: _________ cc. Number of cups tha supply chemical to each seed dump: ________. Total chemical applied to each seed dump: _____________ cc.

17. Number of seed dumps per unit of seed (when recommended treatment rate is expressed as amount applied per 100 kg, per 100 pounds, per bushel, etc.).

Divide treatment rate in cc by cc of chemical per dump:

________ / ________ = ________ (number of seed dumps that are required to apply the specified amount of chemical too the specified amount of seed).

18. Amount of treatment per dump of seed: divide the recommended amount of treatment by the number of seed dumps required to give the required amount of seed:

    ______ / ______ = ______ (number of seed dumps)

19. Seed dump counterweight setting:

    Look in the treater's "Seed Dump Weight Chart/Table" and find the counterweight setting which will give the weight per dump OF THIS CROP SEED as determined above. If the exact weight setting is not shown, extrapolate as required. Enter the extrapolated setting in your table of treatment settings for future use. Check and adjust the setting for accuracy as follows. After setting the seed dump counterweight at the approximate position, slowly feed seed into the treater until ONE SEED DUMP is made. Immediately stop the seed flow. Remove ALL the dumped seed, and weigh it. Adjust the counterweight as required: up for more seed, down for less seed.

    Seed dump counterweight setting: ______________________.

THE TREATING OPERATION:

20. Does the liquid or slurry treatment "foam" in either the main tank or in the metering tank? Does this interfere with the proper application dosage? How can foaming be reduced?

21. Does dust absorb the chemical in the treatment application chamber? Does this present a problem to seed flow or treatment dosage? How can dust be reduced or eliminated?

22. Open the inspection doors to observe the action of the treater's blending mechanism. (CAUTION: KEEP HANDS OUT!) Does this mechanism cause mechanical damage to seed? Does it blend seed and chemical thoroughly? Describe.

23. Select a random sample of 400 treated seed. Examine all sides of each seed under magnification and evaluate the uniformity of treating by classifying the seed as follows:

    a. Number of seed with no visible treatment.

    b. Number of seed with smooth coating of treatment on all sides.

    c. Number of seed with smooth coating of treatment on only one side.

    d. Number of seed with "spots" of treatment on all sides.

    e. Number of seed with "spots" of treatment on only one side.

    f. Are seed being uniformly and acceptably treated? Describe.

24. Is a maintenance manual available, or an operator's manual with a detailed description of recommended maintenance? Is the machine in good condition and well maintained? What changes/improvements in maintenance would improve it?

25. Points, conditions, or elements of special interest.

26. In the notebook that carries your study manual and your own observations, attach a copy of the operation and maintenance manuals of this model of machine that is installed in your facility.

chapter seventy

Bagger-weigher characteristics

Conditioner name:

Date:

Seed conditioning plant/machine location:

Machine brand, model number, other identification:

Year machine installed; condition:

Installation and feeding seed/waste materials to/from the machine description:

Any problems encountered with operation of the machine:

Procedure:

1. Is an operator's manual available? What is the machine designed to do? Is it being used for its designed purpose? If not, how is it being used? Can its use be changed to take better advantage of its design features? Is another machine(s) needed to improve cleaning performance?

2. What are the bagger-weigher's requirements for:

   a. Power?

   b. Seed supply and feed?

   c. Installation and mounting?

   d. Handling discharged seed?

   e. Dust control and air ducting?

3. How is the bagger-weigher mounted?

4. How is seed fed into the bagger-weigher? Is there a complete separately operated shutoff control of seed flow from the bin into the bagger-weigher?

5. Overall dimensions:

   Height

   Width

   Length

   Height, floor to seed feed intake

   Height, floor to good seed discharge

6. What is the minimum and maximum weight of filled bags that is possible when using this machine?

7. How is filled bag weight determined and controlled?

8. How is seed flow into a bag operated? Describe.

9. How is seed flow stopped when the bag contains the desired weight of seed? Describe.

10. How is an empty bag put into the filling position? How is it held on the bagger-weigher?

11. How is the filled bag released from the bagger-weigher?

12. How much bag handling by the operator is required?

13. Points, conditions, or elements of special interest.

14. In the notebook which carries your study manual and your own observations, attach a copy of the operation and maintenance manuals of this model of machine which is installed in your facility.

15. Where is seed handled by this machine coming from, and where is it being fed to?

16. What additional facilities, structures, subsidiary equipment, etc., is required for this machine to operate effectively and efficiently?

chapter seventy-one

Bagger-weigher operation

Conditioner name:

Date:

Seed conditioning plant/machine location:

Machine brand, model number, other identification:

Year machine installed; condition:

Installation and feeding seed/waste materials to/from the machine description:

Any problems encountered with operation of the machine:

Procedure:

1. Is a maintenance manual available, or an operator's manual with a detailed description of recommended maintenance? Is the machine in good condition and well maintained? What changes/improvements in maintenance would improve it?

2. Who is the manufacturer, what is the brand name and model of the bagger-weigher?

3. What is the smallest, and the largest, weight of seed this model can bag?

4. How is the bagger-weigher installed? How is seed fed into it? How can seed flow be stopped/started/controlled?

5. How is the desired weight of seed in the bag set? What is the adjustment procedure?

6. How is seed flow into the bag initiated and then stopped when the desired weight is reached?

7. Set the bagger-weigher for the desired weight, and fill 20 bags, trying to operate at normal working speed. What problems were encountered (in this trial and in other operating times)? How can these problems be alleviated?

8. Close the bags, in the normal manner. Using a portable scale, weigh each of the bags. How many were "off the desired weight?" How much? How can this variation be eliminated?

9. What are the maintenance and requirements of the machine?

10. If the bagger-weigher is in the most commonly used mounting, it is mounted on the discharge outlet of the bagging bin. How is it mounted? Does the bin have a separate discharge shutoff gate, as it should have? Why?

11. Is the bagger-weigher difficult to clean before changing seed lots? What is required?

12. Points, conditions, or elements of special interest.

chapter seventy-two

Bag closer characteristics

Conditioner name:

Date:

Seed conditioning plant/machine location:

Machine brand, model number, other identification:

Year machine installed; condition:

Installation and feeding seed/waste materials to/from the machine description:

Any problems encountered with operation of the machine:

Procedure:

1. What is the manufacturer and model of the bag closer?

2. Is a maintenance manual available, or an operator's manual with a detailed description of recommended maintenance? Is the machine in good condition and well maintained? What changes/improvements in maintenance would improve it?

3. Overall dimensions:

   Height

   Width

   Length

   Height, floor to seed feed intake

   Height, floor to good seed discharge

4. What are the bag closer's requirements for:

   a. Power?

   b. Installation and mounting?

5. How is the bag closer installed? Is it mounted on a bag conveyor belt? How is the mount adjusted, as needed?

6. How is the bag closer used?

7. What are the bag closer's requirements for thread, and how is thread supplied to it?

8. Follow the path of the thread from the spool into the eye of the needle. What does it touch? What is the purpose of each point?

9. Are there problems in the flow of thread? What causes the problems? How can this be remedied?

10. How is the thread cut when the bag is fully closed? Are there problems in this? Is the operator required to do inconvenient actions?

11. How and when/where are tags (and seals, when used) attached to the bags?

12. How are filled bags fed to the bag closer? Is it a simple, efficient movement? Are there problems, inefficiencies, or excessive handling requirements? How can this situation be improved.

13. How are filled bags moved? Is the bag closer mounted on a bag conveyor belt?

14. How are filled/closed bags moved out of the closing site, for storage or delivery? Can this procedure be improved to reduce work strain, labor requirements, or time?

15. Points, conditions, or elements of special interest.

16. In the notebook that carries your study manual and your own observations, attach a copy of the operation and maintenance manuals of this model of machine that is installed in your facility.

chapter seventy-three

Bag closer operation

Conditioner name:

Date:

Seed conditioning plant/machine location:

Machine brand, model number, other identification:

Year machine installed; condition:

Installation and feeding seed/waste materials to/from the machine description:

Any problems encountered with operation of the machine:

Procedure: Using a seed lot, operate the bag closer on 20–25 bags and note the following:

1. Is a maintenance manual available, or an operator's manual with a detailed description of recommended maintenance? Is the machine in good condition and well maintained? What changes/improvements in maintenance would improve it?

2. Fill and close 20–25 bags and evaluate every minute detail of the operation. Can any aspect of the operation be improved?

3. Is the bag closer held by hand, suspended on a wire or cable, or mounted on a stand on a bag conveyor belt?

4. Is it difficult to move the filled bag from the filling station into the start of the bag sewer-closer operation?

5. Is it difficult to feed the unclosed bag into the bag closer and start closing the bag? Why? How can this be alleviated?

6. If the bag closer's needle hits a large seed (such as barley, wheat, or oat), it will often break the needle. How is this prevented?

7. How is the sewing machine thread cut when the bag sewing is completed? It this a problem? Why?

8. How is the filled bag moved from the bag closing station? Is this a problem? How can it be improved?

9. Points, conditions, or elements of special interest.

chapter seventy-four

Determining seed conditioning requirements

Conditioner name:

Date:

Seed conditioning plant/machine location:

Machine brand, model number, other identification:

Year machines installed; condition:

Installation and feeding seed/waste materials to/from the machine description:

Any problems encountered with operation of the machine:

Procedure: Use lab or model equipment and internal quality control testing procedures. Determine conditioning requirements and provide the following information or answers to questions, for each of several seed lots.

1. Crop and variety, seed lot number.

2. Seed condition (not conditioned, partially conditioned, etc.; describe) and appearance.

3. Internal quality control purity analysis

   % Pure seed

   % Impurities

   Name and description of each undesirable material that should be removed

   % Immature, broken, etc., crop seed

4. Impurities that should be removed.

Impurity(ies) to be removed	Physical difference by which it can be separated	Separator(s) that could be used

5. Moisture content

Initial:

Desired:

Drying required:

6. Estimated drying weight loss:

Formula: $\% \text{ weight loss} = \dfrac{(100 - \text{initial } \% \text{ moisture}) \times 100}{(100 - \text{final } \% \text{ moisture})}$

Plus 0.50% of initial weight for invisible loss

7. Estimated conditioning requirements.

No. in sequence	Cleaner or separator	Screens, settings, etc., to be used	Undesirable material(s) to be removed

8. Estimated weight loss in conditioning this lot.

Source	% Loss	Remarks
Drying		
Cleaning (list each machine, with estimated cleaning loss from each)		
Total % weight loss		

9. Expected quality, condition, and appearance of good seed after the above-mentioned conditioning.

10. Draw the complete conditioning flow diagram for this seed lot. Show (beside each machine) screens, cylinder sizes, settings, etc., to be used.

chapter seventy-five

Determining conditioning requirements for a specific separation problem

Conditioner name:

Date:

Seed conditioning plant/machine location:

Machine brand, model number, other identification:

Year machine installed; condition:

Installation and feeding seed/waste materials to/from the machine description:

Any problems encountered with operation of the machine:

Procedure: Evaluate a suitable seed lot or large sample, determine its complete conditioning requirements, and answer the following. Do this for several lots.

1. Crop and variety, seed lot number.

2. Seed condition (not conditioned, partially conditioned, etc.; describe in detail) and appearance.

3. Specific impurity(ies) to be removed.

4. Conditioning possible or required to remove the impurities.

Impurity(ies) to be removed	Physical difference by which each can be separated	Separator(s) that could be used

5. Estimated conditioning requirements.

No. in sequence	Cleaner or separator	Screens, settings, etc., to be used	Undesirable materials removed
1			
2			
3			
4			
5			
6			
7			
8			
9			
10			

6. Evaluate quality, condition, and appearance of good seed after the above-mentioned conditioning. Are changes required in ANY aspect?

7. Draw the complete conditioning flow diagram for this seed lot. Show (beside each machine) screens, cylinder size, etc., to be used.

chapter seventy-six

Determining sequence to set up, adjust, and operate conditioning machines

Conditioner name:

Date:

Seed conditioning plant/machine location:

Machine brand, model number, other identification:

Year machine installed; condition:

Installation and feeding seed/waste materials to/from the machine description:

Any problems encountered with operation of the machine:

Procedure: The sequence of setting up/adjusting/operating the air-screen cleaner is shown as an example. Study adjustment and operation of other separators and prepare a similar sequence of adjustments for each machine. Follow the same procedure for all machines.

Air-screen cleaner adjustment sequence

1. Using internal quality control procedures, analyze a sample of the seed lot; determine what must be separated from the good seed, and the physical differences that may be used to make the required separations.

2. Select screens with the proper perforation sizes and shapes, by experimenting with hand screens.

3. Place the screens in the proper positions in the cleaner; secure the screens; and place all spouts, etc., in their operating positions. Verify proper positioning of screen brushes. Also verify that the necessary elevators, bins, bags, etc., are in position to deliver seed to the bin over the cleaner; receive and move good seed to the next conditioning machine; and handle all waste products discharging from the cleaner.

4. Close the feed on the cleaner's overhead receiving/feed bin so that no seed can go into the cleaner.

5. Fill the bin over the cleaner with the seed lot to be cleaned and arrange seed feeding and supply so that the entire seed lot can be continuously fed into the conditioning operation without interruption or variation in seed feed and flow.

6. Reduce the upper and lower air separation fan/shutoffs so that the air blasts that make the separations are almost closed.

7. Turn the cleaner on, after being sure that all components are in place, properly secured, and operating properly.

8. Open and/or activate the feed mechanism of the cleaner's hopper so that a moderate amount of seed flows across the screens.

9. Adjust the upper air to remove the desired dust and light materials.

10. Adjust the screen shoe vibration to get the most effective movement of seed on and across the screen surfaces, so as to ensure getting the desired separations. Use condition of seed flow as related to the separation that is being made on the most critical separating screen, as the determining factor.

11. Adjust the lower air to remove the light seed and materials that must be removed.

12. Verify and adjust the pitch (inclination) of each screen individually. Increase pitch to move seed over the screen faster. Decrease pitch to keep seed on the screen longer so as to increase the completeness of the desired separation.

13. Improve the capacity and separation of the cleaner, by starting with step 8 and readjusting as required to make improvements in the complete removal of undesirable seed/particles. Often, one or more screens may have to be changed.

chapter seventy-seven

Selected machine adjustment sequence

At the beginning of conditioning a seed lot, each separator used should be set to have the approximate adjustments determined by pre-evaluation of samples as being the best to obtain the desired complete separation. As soon as the cleaning/separating begins, evaluate the separation made by each machine for complete removal of the desired undesirable particles/seed. Make required minor (or major!) changes in adjustments to achieve the complete desired separations.

For each machine consider the following:

1. Sequence of use and initial adjustments.

Sequence	Adjustment or sequence of use
1	
2	
3	
4	
5	
6	
7	
8	
9	

2. Changes in adjustments or use.

Adjustment sequence	Change in adjustment or operation
1	
2	
3	
4	
5	
6	
7	
8	
9	

chapter seventy-eight

Special packaging machines

Conditioner name:

Date:

Seed conditioning plant/machine location:

Machine brand, model number, other identification:

Year machine installed; condition:

Installation and feeding seed/waste materials to/from the machine description:

Any problems encountered with operation of the machine:

Procedure: Although most agricultural crop seed is handled in bags of fairly large size, many crop seeds are packaged in smaller units. Such packaging in small packets is especially common for vegetable and flower seed.

The machines that package seed in small packets are more sensitive than machines that package seed in large bags, and they require careful adjustment, maintenance, and operation to ensure constant accuracy. These operations are usually in a separate installation, commonly in air-conditioned, dehumidified rooms to ensure conditions that do not affect either the seed or the sensitive operations of the equipment.

1. What seed is being packaged in small packets or bags or packages? What size or weight of packages is being used?

2. What are the overall dimensions of the machine?

   Height

   Width

   Length

   Height, floor to seed feed intake

   Height, floor to good seed discharge

3. What are the packager's requirements for

   a. Power?

   b. Installation and mounting?

   c. Special needs?

4. How is the desired seed weight set on the packaging machine?

5. How are empty packages set or fed into the packager?

6. How much operator handling/work/adjustment/oversight is required? Doing what?

7. How are filled packages handled?

8. What are the weak points or points requiring most supervision and control of this machine?

9. What can be done to improve the operation?

Completion of the study program

Upon completion of this study program, the Director (Head of the Seed Center, whether identified as Director or other title) should examine the staff member's Personal Reference Notebook. If it is considered to show in-depth completion of the studies organized in this program, the Director should complete the following Certificate and, with appropriate ceremony, award this **Certificate of Master Seed Conditioner** to the staff member. It is recommended that the awarding of this Certificate be announced on local TV and in newspapers and other news media.

Because the Seed Center's program will improve, in terms of better seed quality, less loss of good seed, reduced operating time and operating costs, and other factors, it is recommended that the efforts of the new **Master Seed Conditioner** also be reflected in his/her salary.

It is hoped that the Seed Center can reap the benefits of having several **Master Seed Conditioners** on the staff.

Certificate of Master Seed Conditioner

This Certificate of **Master Seed Conditioner** is awarded to

____

in recognition of the successful completion of the work and study program established internationally for this title. The recipient is authorized to use the title Master Seed Conditioner in official and unofficial activities, and to receive all the technical and professional recognition and honors associated therewith.

____

Director

____

Seed Center

____

Date of Award